PROPHETIC CYBER INTELLIGENCE

Volume 1 (SECURITY: Guarding the Spiritual Grid)

DR. DWAYNE M. FRASER

Scripture quotations are from the King James Version (KJV) of the Bible, which is in the public domain. Scripture taken from the New King James Version®, NKJV®. Copyright © 1982 by Thomas Nelson. Used by permission. All rights reserved. Scripture quotations are taken from the Holy Bible, New International Version®, NIV®. Copyright © 1973, 1978, 1984, 2011 by Biblica, Inc.™ English Standard Version, ESV®. Wheaton, IL: Crossway Bibles, 2001. The Living Bible, TLB®. Wheaton, IL: Tyndale House Publishers,1971. Used by permission. All rights reserved worldwide.

PROPHETIC CYBER INTELLIGENCE Volume 1
SECURITY: Guarding the Spiritual Grid

Dr. Dwayne M. Fraser
516 Sosebee Farm Rd
#1243
Grayson, GA 30017
www.dwaynefraser.com

ISBN: 978-0-9881981-3-5

Printed in the United States of America
©2024 by Dr. Dwayne M. Fraser
All rights reserved.

No part of this book may be reproduced or transmitted in any form or by any means, electronic or mechanical, including photocopying, recording, or by any information storage and retrieval system, without written permission from the publisher.

Living Water Publications
AREC Global
P.O. Box 1243
Grayson, GA 30017
www.abundantrain.org

Acknowledgments

I would like to extend my deepest gratitude to everyone who made this book possible.

To my wife, Dr. Keri Fraser, thank you for your unwavering support, love, and patience throughout the long hours of writing and revisions. You were my constant source of encouragement. To my wonderful family, including my church family (AREC Global) thank you for never allowing me to let go of this dream.

To my mentors, Apostle R. A. Morton, Dr. Frank Ofosu-Appiah, Apostle John Lewis and Apostle Lynn'Da Threat, your wisdom and guidance have been instrumental in shaping the content and direction of this book. I am forever grateful for your insight and encouragement along the way.

A heartfelt thank you to my editor, Dr. Tameka Scotton, whose keen eye and dedication brought out the best in every chapter. To the entire publishing team at Living Water Publications, thank you for believing in this project and bringing it to life with professionalism and care.

To my colleagues and friends who offered feedback and inspiration - thank you for your contributions. Your advice, insights, and friendship were invaluable throughout this journey.

Lastly, to all the readers and those who have supported my work over the years, this book is for you. Your enthusiasm and belief in this message fueled me from start to finish.

CONTENTS

1. NAVIGATING EMERGING THREATS: Insights from Biblical and Digital Perspectives.................. 7

2. FORTIFYING THE HEART: Protecting Our Inner World from Exploitations........................ 21

3. A SECURE CONNECTION: Building Trust.............. 37

4. ESSENTIAL UPGRADES: Strengthening Resilience for the Future 47

5. THE MAN IN THE MIDDLE – Interception or Intercession....................................... 59

6. EMBRACING CHANGE: The Power of Rebooting Your Life... 73

7. FROM STRUGGLE TO STRENGTH: Navigating the Path of Recovery................................... 87

8. DIVINE FIREWALL: Safeguarded by God............... 95

-1-
NAVIGATING EMERGING THREATS
(Insights from Biblical and Digital Perspectives)

Emerging Threats

Imagine logging into your bank account while on vacation. You expect to see the usual balance, but instead, you discover transactions you never made. Your heart races as you realize your savings account has been drained. You scramble to check your other accounts (i.e. email, social media) and you notice strange activity everywhere. Panic sets in as you grasp the reality that your personal information has been stolen. As you report the theft, a cybersecurity expert explains how these emerging threats are becoming more sophisticated and targeting everyday people like you. *It's not just about hacking anymore*, the expert says. *They're using new tactics to trick you and get past your defenses.*

This experience is a harsh reminder that as technology evolves, so do the threats. It's crucial to stay vigilant and to understand that what worked yesterday might not protect you today. Just as we need to be aware of these new dangers (to guard our personal information), God wants to protect us from the unseen threats in our lives. He wants to guide us through ever-changing challenges with divine wisdom.

In today's rapidly evolving world, we are continually confronted with emerging threats that challenge our security, stability, and well-being. They range from sophisticated cyber-attacks that target our digital infrastructures to societal disturbances and personal vulnerabilities. These threats demand awareness and proactive measures. While these challenges may seem unprecedented, they often reflect deeper, enduring struggles that have been addressed in biblical and prophetic writings. By drawing parallels between modern issues and scriptural insights, we can improve our understanding of how to navigate and reduce the risk of these threats. This, in turn, will ensure our physical and spiritual resilience. This perspective is particularly relevant when considering emerging threats.

Emerging threats are new or evolving dangers that can harm digital systems, individuals, organizations, or society on a whole. These threats are often unfamiliar and can come from various sources, including technology, nature, or human behavior. Because they are new or constantly changing, they can be particularly challenging to understand and address. Imagine a new type of virus that spreads quickly and affects a lot of people before doctors know how to treat it. This is an example of an emerging threat in the health sector.

In the world of technology, emerging threats often refer to new types of cyberattacks. As technology advances, so do the methods used by cybercriminals to exploit vulnerabilities in computer systems. For instance, a new kind of malware might be developed to steal personal information or disrupt critical services. Because these threats are new, security defenses may not be in place. Therefore, the threats are dangerous.

Emerging threats can also come from environmental changes. For example, climate change can lead to more frequent and severe natural disasters (i.e. hurricanes, wildfires, and floods). These environmental threats can have a significant impact on communities, disrupt lives, and cause widespread damage. As the climate continues to change, new patterns of extreme weather and environmental stress can pose emerging threats that require new approaches to manage.

Understanding and addressing emerging threats requires staying informed and being proactive. This means keeping up with the latest information about potential dangers and taking steps to protect oneself and one's community. Being aware and ready for emerging threats can help mitigate their impact and ensure a safer future. Some suggested methods of preparation include: (1) installing updated security software to guard against cyber threats, (2) preparing for natural disasters by having an emergency plan, or (3) supporting policies that address climate change.

The Bible is a source of timeless wisdom. It provides profound awareness into the nature of emerging threats that contribute to human challenges. The Bible also provides insight to the principles required to overcome these challenges. Prophetic scriptures, in particular, offer foresight into the varied dimensions of calamities that humanity faces. The Bible is not just a book of ancient texts. It is also a credible source that provides guidance on how to maintain faith, integrity, and resiliency in the face of adversity. By examining emerging threats through the lens of biblical and prophetic insights, we not only identify the underlying challenges, but also find inspiring principles to guide us. This integrated approach allows us to address current issues with wisdom and foresight. It also ensures that we protect our digital environments and secure our hearts and minds in accordance with Divine principles.

Malicious Threats
Emerging threats can come in many forms and have wide-ranging impacts. In the area of cybersecurity, an emerging threat might be

a new type of malicious software known as malware. It is designed to infiltrate, damage, or disable computer systems, networks, or devices. Its primary intent is to steal sensitive information, corrupt data, or exploit system vulnerabilities for malicious purposes. This digital menace parallels the spiritual threats highlighted in the Bible. For example, where sin and deceit seek to corrupt and steal the purity of the heart and mind.

In the digital world, malware operates covertly, embeds itself within systems, and executes harmful tasks without the user's knowledge. Likewise, our adversary (the devil) works subtly to infiltrate people's lives and lead individuals astray. The Bible speaks to this in John 10:10 (NKJV): *"The thief does not come except to steal, and to kill, and to destroy. I have come that they may have life, and that they may have it more abundantly."* Just as malware aims to steal, kill, and destroy digital confidentiality, integrity and availability, deceit seeks to undermine our spiritual well-being.

The deceptive nature of malware can be compared to the deceitfulness of sin. Let us look at Jeremiah 17:9 (NKJV). *" The heart is deceitful above all things, and desperately wicked; who can know it?"* This scripture highlights the sinister nature of sin. Like malware, it can corrupt and damage the integrity of the heart. Malware often disguises itself as legitimate software. Sin can appear attractive or harmless. It can also mask its true intent to harm and lead individuals away from righteousness.

To safeguard against malware, we must employ robust cybersecurity measures. For example, antivirus software, firewalls, regular updates, and vigilance against suspicious activities. Similarly, the Bible provides guidance on protecting our hearts and minds from spiritual corruption, which requires the following:

1. **Regular Self-Examination and Repentance:** Just as systems require regular scanning to detect and remove malware, we need to examine our lives for hidden sins and repent. Lamentations 3:40 (NIV) advises, *"Let us examine our ways and test them, and let us return to the Lord."* 1 John 1:9 (KJV) says, *"If we confess our sins, he is faithful and just to forgive us our sins, and to cleanse us from all unrighteousness."*

2. **Guard Against Deception:** Proverbs 4:23 (NIV) instructs, *"Above all else, guard your heart, for everything you do flows from it."* Protecting our hearts from deceit and malicious influences is like utilizing a firewall in cybersecurity. The firewall filters out harmful data and prevents unauthorized access.

3. **Seek Wisdom and Discernment:** James 1:5 (NIV) says, *"If any of you lacks wisdom, you should ask God, who gives generously to all without finding fault, and it will be given to you."* Wisdom acts as an antivirus program. It helps to identify and reject harmful influences before they take root in our lives.

4. **The Armor of God:** Ephesians 6:11 (NIV) directs us to, *"Put on the full armor of God, so that you can take your stand against the devil's schemes."* This spiritual armor includes truth, righteousness, faith, salvation, and the Word of God. Together, they form an impenetrable barrier and defense against spiritual malware.

5. **Watch and Pray:** In Matthew 26:41, we are advised to *"Watch and pray so that you will not fall into temptation. The spirit is willing, but the flesh is weak."* Being aware and prayerful of spiritual dangers helps keep us prepared and protected. In the same manner, we must maintain updated security awareness and protocols in a digital environment.

Understanding malware as an emerging threat provides a tangible analogy for the spiritual dangers that threaten our hearts and minds. By relying on biblical principles, we can adopt a proactive stance against these threats and ensure both our digital and spiritual resilience. Just as robust cybersecurity measures protect against malware, spiritual disciplines safeguard our hearts. They also enable us to live lives of integrity and faithfulness in an increasingly changing and complex world.

Viral Threats

Viral diseases represent another significant emerging threat in today's world. This kind of threat impacts public health and global economies with unprecedented speed and severity. The rapid spread of infectious diseases, such as COVID-19, demonstrates the potential for viruses to disrupt societies and overwhelm healthcare systems. Understanding this threat through a biblical lens provides insight into how we must approach such challenges with faith and wisdom.

Viral diseases can spread quickly and unpredictably, often outpacing the capacity of healthcare systems to respond effectively. The COVID-19 pandemic, for example, illustrated how a new virus can lead to widespread illness, economic disruption, and social upheaval. This rapid and often uncontrollable virus parallels the broader concept of pestilence and disease described in the Bible. It also highlights the need for preparedness and resilience.

The Bible addresses the concept of disease and pestilence in several ways, which reflects both immediate and prophetic concerns. In Deuteronomy 7:15 (NKJV), it is written, *"And the Lord will take away from you all sickness, and will afflict you with none of the terrible diseases of Egypt which you have known, but will lay them on all those who hate you."* This verse emphasizes the idea that while disease can be a part of human experience, divine protection is promised for those who are faithful to God.

Another significant prophetic reference can be found in the book of Revelation. Revelation 6:8 (KJV) mentions, *"And I looked, and behold a pale horse: and his name that sat on him was Death, and Hell followed with him. And power was given unto them over the fourth part of the earth, to kill with sword, and with hunger, and with death, and with the beasts of the earth."* This scripture points to the devastating impact of various calamities (i.e. disease, famine, and plague) that will prophetically occur at a certain time.

The Bible's reflections on disease and pestilence encourage preparedness and response both spiritually and naturally. Preparing for viral threats (in the natural world) involves maintaining good health practices, supporting public health measures, and promoting community awareness. However, preparing for viral threats (in the spiritual world), calls for prayer, faith, and shared support. In other words, we must align ourselves with the biblical principle of seeking God's protection and guidance.

The emerging threat of viral diseases underscores the importance of understanding both the practical and spiritual dimensions of health and crisis management. By reflecting on biblical and prophetic insights, we gain valuable perspectives on how to navigate these challenges with faith, wisdom, and determination. Embracing these concepts can help us prepare for and respond to viral threats. This will also ensure our physical safety and spiritual well-being in times of uncertainty.

Cyber Terrorism Threats

Cyber terrorism is a growing concern that combines elements of both cybersecurity and terrorism. Cyber terrorists use digital tools (i.e. malware, phishing, advanced persistent threats, and social engineering) to carry out attacks that can cause widespread disruption and fear. Cyber terrorists aim to disrupt, damage, or exploit critical infrastructures (i.e. power grids, financial systems, and communication networks). Their intent is to cause widespread panic, economic turmoil, and social instability. These attacks can be difficult to detect and identify, which makes them dangerous. As technological advancements continue, the methods and capabilities of cyber terrorists also evolve and pose significant challenges to national and global security.

While the Bible may not directly address cyber terrorism, it offers timeless principles that can be applied to this emerging and modern threat. One such principle is the call for vigilance and preparedness. In 1 Peter 5:8 (NIV), it is written that we should, *"Be alert and of sober mind. Your enemy the devil prowls around like a roaring lion looking for someone to devour."* This verse stresses the necessity of being constantly watchful and ready to respond to threats. Just as believers are urged to stay on the alert against spiritual adversaries, individuals and organizations must remain alert to the potential dangers posed by cyber terrorists. They must also take proactive measures to secure their digital environments.

Prophetic guidance from the Bible emphasizes the importance of wisdom, integrity, and joint support in facing and overcoming threats. Proverbs 2:11 (NIV) states, *"Discretion will protect you, and understanding will guard you."* This passage focuses on the protective power of discernment and wisdom. It also encourages the application of these qualities in safeguarding against spiritual attacks. Additionally, the principle of shared security and support is reflected in Ecclesiastes 4:12 (NIV): *"Though one may be overpowered, two can defend themselves. A cord of three strands is not quickly broken."* This verse speaks to collaboration and shared support, which are essential in developing resilient defenses against cyber threats. By fostering a culture of cooperation and shared responsibility, communities and nations can better prepare for and mitigate the impact of cyber terrorism.

Cyber terrorism is a significant and emerging threat that requires awareness and proactive measures. Biblical principles of watchfulness, wisdom, and shared responsibility offer valuable guidance in addressing this modern danger. By staying alert, seeking divine wisdom, and working together, we can build robust defenses against cyber terrorism. This will ensure the security and stability of our digital and physical worlds.

Artificial Intelligence Threats
Artificial Intelligence (AI) represents a profound and emerging threat with far-reaching implications for society. As AI technologies rapidly advance, they introduce new challenges and uncertainties (i.e. issues of control, ethics, and potential misuse). Exploring these challenges through a biblical lens offers valuable

insights into the moral and spiritual considerations that surround technological advancement.

AI systems have the potential to transform various aspects of life (ranging from healthcare and finance to security and entertainment). However, their capabilities also pose significant risks. These include concerns about privacy, the potential for job displacement, the ethical use of AI in decision-making, and the possibility of autonomous systems making harmful decisions. The rapid development of AI introduces complexities that require careful consideration of its impact on humanity and society.

From a Biblical perspective, we can draw from the foundational principles (e.g. knowledge, power, and human responsibility) that can be applied to the context of advanced technologies such as AI. Let us look at the construction of the Tower of Babel. Genesis 11:6 (NKJV) states, *"And the Lord said, 'Indeed the people are one and they all have one language, and this is what they begin to do; now nothing that they propose to do will be withheld from them'"*. This passage focuses on the potential of human innovation, as well as the dangers of pursuing knowledge and power without God's guidance or moral considerations.

The pursuit of knowledge and the development of new technologies are not essentially negative, but they come with responsibilities. Proverbs 1:7 (NIV) states, *"The fear of the LORD is the beginning of knowledge, but fools despise wisdom and instruction."* This verse highlights the importance of approaching advancements with humility and reverence for divine wisdom. It also suggests that knowledge should be pursued in alignment with ethical and moral principles.

Additionally, AI's potential for misuse and harm can be compared to the concept of false prophets and deceptive practices, which the Bible warns against. In Matthew 7:15 (NIV), Jesus said, *"Watch out for false prophets. They come to you in sheep's clothing, but inwardly they are ferocious wolves."* This metaphor focuses on our need to be aware and discerning in the face of potential threats, including the misuse of advanced technologies.

Addressing the challenges posed by AI requires a combination of ethical foresight, responsible innovation, and robust governance. The Bible encourages believers to seek wisdom and understanding in all endeavors. Proverbs 3:5-6 (NIV) advises us to, *"⁵Trust in the LORD with all your heart and lean not on your own understanding; ⁶in all your ways submit to him, and he will make your paths straight."* This scripture emphasizes the importance of seeking God's direction. In the same manner, we should align technological advancements with ethical and spiritual principles.

The emerging threat of AI not only brings opportunities and challenges. It also requires a careful and ethical approach to its development and implementation. Biblical principles on knowledge, ethical and moral behavior, and watchfulness offer valuable insights for navigating these intricacies. By applying these principles, we can ensure that the advancements in AI contribute positively to humanity. This measure will also safeguard against potential risks and maintain alignment with ethical, moral, and spiritual values. Embracing a perspective that integrates faith and ethics into technological progress can help us address the emerging threats of AI with wisdom and responsibility.

In navigating the complex landscape of emerging threats, we recognize the profound relationship between digital advancements and the timeless wisdom found in the Bible. From constant defense against sophisticated cyberattacks to global combat against viral diseases, the principles of proactive defense and community resilience are vital. Cyber terrorism has the capacity to disrupt and instill fear. Therefore, it demands coordinated responses that mirror the biblical call to stand firm against evil. Organized crime's adaptation to the digital age necessitates advanced strategies and reflects the need for wisdom and discernment in all aspects of life. While presenting both opportunities and risks, the rise of artificial intelligence emphasizes the importance of ethical considerations and moral guidance. In the same manner, the Bible exhorts us to seek after righteousness.

By understanding and addressing these emerging threats, we draw on advanced knowledge and biblical principles that safeguard our physical, digital, and spiritual environments. The Bible's prophetic insights and ethical guidelines offer a roadmap for navigating the challenges of today's world. They also remind us of the enduring values of integrity, watchfulness, and kindness. Integrating these principles will help ensure a safer, more resilient future that is rooted in the wisdom of the ages and adapted to the realities of our time. Through this holistic approach, we can protect and nurture our communities. Doing so will foster a world where technological progress and spiritual well-being thrive together.

Inspirational Downloads

-2-
FORTIFYING THE HEART
(Protecting Our Inner World from Exploitations)

> **Exploits**
>
> Imagine you are the CEO of a successful company. After a long day, you sit down to check your email. Suddenly, an urgent message catches your eye. It looks like it is from the head of your IT department. It is a warning of a security breach that needs immediate attention. The email urges you to log in and verify the company's credentials to protect sensitive data. Without thinking twice, you click the link and enter the information. However, later, you discover that your company's most confidential data has been leaked and your financial accounts have been compromised. The panic becomes overwhelming as you realize the breach started with one email. You call the IT department. An expert tells you, *"They used a spear-phishing exploit tailored just for you (the CEO) because they knew you held the keys to everything."* This scenario is a stark reminder that no one is immune to these sophisticated tactics, especially no one in leadership. Just as we need to guard against these exploits in business, God calls us to seek His wisdom. In doing so, we can protect ourselves from the strategies that aim to misuse our power, exploit vulnerabilities, and lead us astray.

The term *exploit* carries significant weight in the realm of information systems security. An exploit is a software tool that is designed to manipulate decision-makers, turn authority into a vulnerability, and take advantage of a flaw in a computer system (for example, by installing malware). This term incorporates various types of malicious software that are designed to infiltrate and damage systems.

Exploits mirror a broader, more sinister concept found throughout human history and even within biblical teachings: *exploitation*. To exploit is to benefit unfairly from the work of others, typically by overworking or underpaying them. It involves abusing and manipulating people for selfish gain. Similarly, in cybersecurity, exploitation involves breaching secure networks (or information systems) in violation of established policies, which leads to corruption and misuse.

The Bible provides numerous examples of exploitation and the moral consequences that follow. In Isaiah 10:1-2 (NIV), the prophet warns, *"¹Woe to those who make unjust laws, to those who issue oppressive decrees, ²to deprive the poor of their rights and withhold justice from the oppressed of my people, making widows their prey and robbing the fatherless."* This condemnation of unjust practices is like the condemnation of cyber exploitation, where malicious actors take advantage of vulnerabilities to cause harm. The ethical implications of both forms of exploitation are clear: They represent a fundamental violation of trust and justice. From a cybersecurity perspective, exploitation is often achieved through the deployment of malware.

Just as the Bible cautions against the exploitation of the vulnerable, it is crucial for individuals and organizations to protect themselves against cyber exploitation. This protection involves installing security software that manages access control, provides data protection, and secures systems against viruses, malware, and network intrusions. Proverbs 4:23 (NIV) advises, *"Above all else, guard your heart, for everything you do flows from it."*

Likewise, guarding our digital devices and systems is essential for maintaining their integrity and functionality.

Understanding the parallels between biblical exploitation and Information Systems Security exploitation helps us grasp the ethical and practical importance of watchfulness and protection. By recognizing the manipulative tactics used in both spheres, we can be better prepared to defend ourselves against them. Whether through spiritual guidance or technical safeguards, the principle remains the same: watchfulness and proactive measures are essential to prevent exploitation and ensure integrity and security.

As we delve deeper into the topic of exploitation, we will investigate how exploits (in information systems security) can be identified, mitigated, and prevented. We will also draw further parallels with biblical teachings that emphasize the ethical requirements needed to guide us. Just as biblical teachings encourage us to stand against exploitation and protect the vulnerable, we must also stand against cyber threats and protect our digital assets. By doing so, we uphold principles of integrity and security in both our spiritual and digital lives.

Understanding that vulnerabilities exist, an enemy, hacker, or malicious actor will exploit weak areas in our computer systems, networks, or digital devices. This parallels how our lives can be exploited in our walk with God. The enemy identifies a weakness, and that vulnerability becomes the target. Strengthening the vulnerable areas of our lives is crucial to prevent the enemy from exploiting those weaknesses. Understanding these dynamics sets the stage for our discussion on how to guard against exploitation in both the digital and spiritual realms.

BIBLE PRINCIPLES AND SECURITY CONCEPTS

WHAT AFFECTS THE HEART

Let us dive into the influences of the enemy, wickedness, evil, and sin. We will examine how they affect the heart. Understanding these influences can be effectively illustrated through the lens of information security concepts. Jeremiah 17:9 (NIV) states, *"The heart is deceitful above all things and beyond cure. Who can understand it?"* This reflects the risks of social engineering attacks in cybersecurity. Attackers manipulate users to gain unauthorized access, just as sin deceives and corrupts the heart. Matthew 6:15 (NIV) warns, *"But if you do not forgive others their sins, your Father will not forgive your sins"*. This highlights how unforgiveness can lead to emotional and spiritual vulnerabilities, much like data corruption can compromise an entire system.

Genesis 6:5 (NKJV) describes the wickedness of the heart by stating, *"Then the Lord saw that the wickedness of man was great in the earth, and that every intent of the thoughts of his heart was only evil continually"*. This parallels the impact of insider threats in information security. Threats from within can cause significant damage, similar to how wicked intentions from within lead to destructive outcomes. 1 Samuel 15:23 (NKJV) states, *"For rebellion is as the sin of witchcraft, and stubbornness is as iniquity and idolatry"*. The consequences of rebellion are like security policy violations. Ignoring these policies can lead to breaches, just as rebellion against divine commands results in serious repercussions.

Proverbs 6:18 (NIV) discusses, *"A heart that devises wicked schemes, feet that are quick to rush into evil"*. This can be compared to malware. It is designed to cause harm and disrupt systems, as evil thoughts and intentions can corrupt the heart. James 1:14 (NIV) explains, *"but each person is tempted when they are dragged away by their own evil desire and enticed"*. This resembles the risks of phishing attacks, which use deceitful emails or seductive temptations to lure individuals into compromising situations.

The Bible highlights false teachings and deceit in 2 Peter 2:1 (NIV), *"But there were also false prophets among the people, just as there will be false teachers among you. They will secretly introduce destructive heresies, even denying the sovereign Lord who bought them—bringing swift destruction on themselves"*. This is comparable to spoofing attacks in cybersecurity (where systems or users are tricked into accepting false information). False teachings mislead individuals as well. Romans 6:23 (NKJV) addresses the impact of sin on one's spiritual health by stating, *"For the wages of sin is death, but the gift of God is eternal life in Christ Jesus our Lord"*. This parallels system vulnerabilities arising from unpatched software. Just as outdated software can be exploited, unchecked sin leads to spiritual decay and eventual separation from God.

By understanding these comparisons, we can better appreciate the need for vigilance and proactive measures in our spiritual lives. We must also be vigilant in protecting our digital environments. It is equally crucial to guard our hearts against exploitation and harm.

GUARD THE HEART
Now, we will look at how to guard the heart. This is key to avoiding exploitation in our lives. Again, we are admonished from Proverbs 4:23 (NIV), *"Above all else, guard your heart, for everything you do*

flows from it". In the latter part of this scripture, the King James Version of the Bible says, *"for out of it are the issues of life"*. The focus of our adversary is to target the heart. If the enemy bypasses the defenses set to protect that area, the enemy will also influence the behavior of the person that is affected. This is the reason why we must shield our heart.

There are spiritual and biological aspects to the importance of the heart. Spiritually, the most important aspect of the heart is that everything about us flows from it. Whatever we speak, comes from the heart. Whatever we do, comes from the heart. The heart is the source of life. It influences thoughts, actions, and overall well-being. Biologically, the most important function of the heart is pumping blood throughout the body. This ensures that oxygenated blood reaches all tissues and organs. The blood provides them with the oxygen and nutrients they need to function properly, while removing carbon dioxide and other waste products. This continuous circulation is essential for sustaining life and maintaining overall bodily functions. The life of the body is in the blood and the heart is the center for blood distribution and processing.

Understanding the important roles of the heart is essential in knowing why it is just as important for us to safeguard it. In information systems security, protecting or safeguarding a system that stores sensitive and valuable information is critical. Organizations invest as many resources as necessary and go to great lengths to secure the confidentiality, integrity, and privacy of this information. Likewise, we must secure our hearts with the same defensive posture.

Securing the heart is like securing an information system. It requires vigilance and proactive measures. Philippians 4:7 (NIV) tells us, *"And the peace of God, which transcends all understanding, will guard your hearts and your minds in Christ Jesus"*. In this analogy, love acts as the firewall. It provides the primary defense for our hearts, while peace serves as the set of rules that allow or deny influences. Just as a firewall blocks unauthorized access to protect a network, love fortifies our hearts against conflicts and deceptive influences. A firewall in information systems security uses rules to control the flow of traffic, permitting or denying access based on predefined criteria. Similarly, the peace of God sets the guidelines that govern what affects our hearts. It also ensures only positive and nurturing influences are allowed.

Romans 5:5 (KJV) says, *"And hope maketh not ashamed; because the love of God is shed abroad in our hearts by the Holy Ghost which is given unto us"*. Colossians 3:15-16 (KJV) further encourages us by stating, *"[15]And let the peace of God rule in your hearts, to which also ye are called in one body; and be ye thankful. [16]Let the word of Christ dwell in you richly in all wisdom; teaching and admonishing one another in psalms and hymns and spiritual songs, singing with grace in your hearts to the Lord"*. By establishing love as our firewall and allowing peace to govern our interactions, we create a robust defense for our hearts. This is much like a well-configured firewall, which protects a computer system.

Regular self-examination is a vital practice for maintaining spiritual health. In the same manner, vulnerability scanning is essential for securing information systems. Lamentations 3:40 (NIV) advises, *"Let us examine our ways and test them, and let us return to the Lord"*. This highlights the need for introspection and repentance. In information security, vulnerability scanning identifies and addresses potential weaknesses in a system before they can be exploited. Similarly, self-examination helps us uncover areas where we might be susceptible to spiritual or moral weaknesses. By identifying these vulnerabilities, we can take corrective actions to strengthen our character. This process involves deep reflection on our thoughts, behaviors, and relationships. It requires honesty and humility to acknowledge our shortcomings.

Engaging in regular self-examination maintains spiritual vigilance, which guards our hearts and minds against negative influences. Additionally, it fosters growth and transformation. Addressing vulnerabilities in a system improves its security, just as acknowledging our weaknesses enhances our spiritual resilience. It allows us to develop a deeper understanding of ourselves and our relationship with God. Regular self-examination is much like conducting vulnerability scans. Both techniques are preventive measures that are designed to identify and address weaknesses before they cause harm.

Maintaining integrity is crucial, as highlighted in Proverbs 11:3, which states, *"The integrity of the upright guides them, but the unfaithful are destroyed by their duplicity"*. This principle parallels the function of access control in cybersecurity. Integrity ensures that only those with proper authorization can affect the heart. This is similar to how access control systems regulate entry to sensitive data. Avoiding negative influences (as advised in 1 Corinthians 15:33), resembles the role of malware protection.

Malware protection shields systems from malicious software, damage, and threats. In the same manner, we must avoid negative influences and preserve the purity of our hearts. This requires being mindful of the company we keep and the content we consume. By surrounding ourselves with positive influences, we can guard our hearts against corruption. Regularly evaluating our relationships and habits helps maintain this protective barrier. Just as cybersecurity measures need constant updating, our vigilance in maintaining integrity and avoiding negativity must be ongoing. Therefore, integrity and avoidance of negative influences work together to safeguard our spiritual well-being, much like access control and malware protection secure information systems.

Ephesians 4:32 (NIV) encourages us to *"Be kind and compassionate to one another, forgiving each other, just as in Christ God forgave you"*. Forgiveness can be compared to patch management in information systems. Just as security patches address vulnerabilities and prevent potential exploits, forgiveness helps resolve relational issues and maintains spiritual health.

James 1:5 (NIV), says, *"If any of you lacks wisdom, you should ask God, who gives generously to all without finding fault, and it will be given to you"*. Cultivating wisdom is like security awareness training. This training equips users with the knowledge needed to protect systems. Wisdom guides us in making sound decisions and avoiding spiritual pitfalls. Both forgiveness and wisdom are essential for maintaining a healthy and secure spiritual life. Forgiveness heals wounds and restores relationships (like patches fix security holes). Wisdom provides the discernment needed to navigate life's challenges safely. Together, forgiveness and wisdom ensure that our spiritual and relational defenses remain robust and effective. Patch management and security training play a similar role. They keep information systems secure.

1 Peter 5:8 (KJV) states, *"Be sober, be vigilant; because your adversary the devil, as a roaring lion, walketh about, seeking whom he may devour"*. Staying vigilant mirrors the role of an intrusion detection system (IDS). An IDS monitors and responds to suspicious activities. This is similar to how being spiritually alert helps us identify and address potential threats. Romans 12:2 (KJV) says, *"And be not conformed to this world: but be ye transformed by the renewing of your mind, that ye may prove what is that good, and acceptable, and perfect, will of God"*. Renewing the mind is like encryption. Just as encryption transforms and secures data to prevent unauthorized access, renewing the mind protects our thoughts and actions from harmful influences.

In Matthew 7:24 (KJV), Jesus teaches, *"Therefore whosoever heareth these sayings of mine, and doeth them, I will liken him unto a wise man, which built his house upon a rock"*. Building strong foundations parallels system hardening, which strengthens defenses against attacks and ensures stability. Building a strong spiritual foundation safeguards against moral and spiritual instability. Together, vigilance, mental renewal, and foundational strength maintain our spiritual integrity and our security. As a result, these reflect the essential elements of both faith and cybersecurity.

Finally, Proverbs 27:17 (KJV) states, *"Iron sharpeneth iron; so a man sharpeneth the countenance of his friend"*. Seeking accountability resembles the practice of maintaining audit trails in security systems. Audit trails are crucial for monitoring activities and ensuring adherence to security policies. They provide a record of actions and changes that can be reviewed for compliance. Accountability involves mutual support and oversight, which helps us stay true to our spiritual commitments and maintain our personal integrity. Just as audit trails ensure that security measures are followed and any deviations are addressed, accountability fosters a supportive environment where we can remain vigilant and aligned with our values. By applying these protective measures, we not only enhance our defenses. We also unlock the profound rewards that come from nurturing a healthy heart.

REWARDS OF A HEALTHY HEART

Guarding the heart offers significant benefits for our emotional and spiritual well-being. Maintaining a healthy heart positively impacts our mental health, emotional stability, and spiritual growth. These rewards highlight the importance of protecting and nurturing our heart for overall wellness.

Spiritually Rewarding

From a spiritual standpoint, guarding the heart is a crucial aspect of maintaining one's faith and spiritual integrity. First of all, guarding the heart involves mindfulness of thoughts and intentions, which ensures that our thoughts align with positive and godly principles. Secondly, guarding the heart requires discernment in relationships. In the discernment of relationships, we choose to surround ourselves with individuals, who support and encourage spiritual and emotional growth. Lastly, we guard the heart through prayer and meditation. These spiritual practices empower us to foster a deeper connection with God and provide strength against temptations and negative influences.

Emotionally Rewarding
Guarding the heart emotionally is vital for maintaining overall well-being and resilience. This can be achieved through setting healthy boundaries in relationships to prevent emotional manipulation or abuse. Engaging in self-care practices, such as exercise, hobbies, and spending time with loved ones promote mental and emotional well-being. During challenging times, seeking support from friends, family, or professionals provides a necessary safety net and aids in navigating emotional difficulties. Additionally, applying cognitive behavioral techniques to reshape negative thoughts, managing stress effectively, and developing emotional management skills are crucial for maintaining emotional stability. Regular reflection on personal experiences and feelings aids in understanding and addressing emotional triggers. Together, these strategies create a robust framework for maintaining mental health, emotional resilience, and protecting the heart from harm.

Practical Steps
Incorporating practical steps to guard the heart can enhance overall well-being. This includes making healthy lifestyle choices (e.g. eating a balanced diet, getting regular exercise, and ensuring adequate sleep). Other practical steps include practicing mindfulness, meditation, or relaxation techniques to reduce stress and enhance emotional clarity. Additionally, engaging in continuous learning, through activities that promote personal and spiritual growth, (i.e. reading, attending workshops, or participating in community activities), contributes significantly to emotional and mental resilience.

In conclusion, guarding the heart is a multifaceted approach that involves spiritual and emotional strategies to protect and nurture one's inner well-being. Mindfulness of the influences around us and taking proactive steps to care for our hearts can safeguard us from internal and external threats. In return, this can lead to a more fulfilling and balanced life.

Careful attention to our emotional and spiritual health mirrors the vigilance required in information systems security. This is needed to prevent exploitation and misuse. Just as we protect our digital systems from malicious exploits (by implementing robust security measures), we must also fortify our hearts against the influences that seek to undermine our well-being. By doing so, we enhance our emotional stability and spiritual growth, and also achieve a deeper sense of fulfillment and resilience in our daily lives.

Inspirational Downloads

-3-
A SECURE CONNECTION
(Building Trust)

> **Secure Connection**
>
> Imagine a plane soaring through the sky, constantly in communication with air traffic control. The plane's systems transmit real-time data about its location, speed, and altitude, while air traffic control provides up-to-date instructions and guidance. This continuous exchange ensures that the plane is safely on course. The exchange also ensures that the plane will avoid collisions and navigate through changing conditions. This scenario mirrors how secure connections work in technology. Constant, reliable communication is essential.
>
> Our connection with God operates with security. Just as air traffic control ensures the plane's data is accurate and uninterrupted, God maintains a direct and secure line of communication with us. We can confidently share our thoughts and concerns through prayer, knowing that our messages are received and understood. God provides us with timely instructions and clarity. He ensures that we stay on the right path. Because we can hear Him clearly and follow His guidance, we are able to navigate life's journey and arrive at our destination safely.

In this chapter, we will explore the concept of a secure connection. Just as encrypted protocols protect our data from interception in the digital realm, when we pray to God, our communication is shielded from any interception. As a result, we can discover the profound assurance that comes from knowing our connection with God remains secure. It also offers a sanctuary of intimacy and

A Secure Connection

protection. Let us explore the mysteries of encrypted protocols and prayer. During this time, we will unveil the extraordinary depth of our secure connection with God.

A secure connection is encrypted by one or more security protocols. This is to guarantee the security of data flowing between two or more connections (nodes). Data can be transmitted over a cable or wireless connection. When a connection is not encrypted, criminals can easily tap into it. The connection is also prone to threats by malicious software and unexpected events. Anyone, who wants to get information from a non-secured connection can easily do so by compromising the computer's network. Unfortunately, this will allow important data (i.e. logins, passwords and other private information) to be stolen.

Secure connections allow the safe transfer of data from one device to another. In addition to that, secure connections also:

1. Prevent the interception of confidential data.
2. Validate the identity of the source attempting to access and exchange information.
3. Protect information from being viewed or altered by intercepting parties.

To create a secure connection, the source device and the destination device use secure protocols. For clarification, these are a set of encryption rules that both devices understand and can interpret because of the trust relationship that was established between them.

Our relationship to God (through Christ Jesus) involves a secure connection, which is established by trust. God is the source, and we are the destination. When communicating with God, it is important to know that there is an encrypted connection to Him through the Spirit. Because a trust relationship has been established through the Spirit, we can hear God's voice and recognize what He is saying.

Our Connection to God

Our connection to God is predicated on our ability to hear His voice. Prayer connects us to Him. Therefore, He hears us when we pray. However, we must be willing to hear Him when He speaks to us. God disseminates His wisdom and knowledge through prayer. His knowledge is secure information, and we must identify (decrypt) His voice to access His intent and wishes. This is why hearing God's voice and maintaining a healthy and strong relationship with Him is so important.

John 10:27-28 (KJV) shows us the connection between the Good Shepherd (Jesus) and His sheep (believers): "*[27]My sheep hear my voice, and I know them, and they follow me: [28]And I give unto them eternal life; and they shall never perish, neither shall any man pluck them out of my hand*". In this passage of scripture, we see a relationship—the connection between Jesus and His sheep. In a relationship with Jesus Christ, we, His people (His sheep), hear His voice. The key to any connection or relationship is communication. Communication strengthens the bond of the connection and brings that relationship into an even safer space. Hearing Jesus' voice helps us to build a more secure relationship with Him.

The Four Points of Intimacy

John 10:27 (KJV) not only shows that we (Jesus' sheep) hear His voice, but also that He knows us. To know is to comprehend, recognize, or understand. This speaks to the intimacy of the relationship between Jesus and His people. By knowing us intimately, Jesus impacts us in four ways: spiritually, intellectually, emotionally, and physically. In a secure connection, only the parties involved know what is being communicated. Jesus knows everything about our lives. Because of this, He leads us as our Good Shepherd. When we consider the four points of intimacy that Jesus impacts, we can reference the fourth letter of the Hebrew alphabet, Dālet. This is the word for door or gate. Dālet shows us that in our humility as sheep, Jesus grants us access to enter. He is the door that leads to salvation, life, and the power of the One— the One source of all creation and existence, who is none other than God. This is how we come to know God, through Jesus Christ, our Lord and Savior.

As Jesus' sheep, He knows us; we hear His voice; and we follow Him. We will not follow a stranger's voice. This is our experience when we are in a secure relationship or connection with Jesus. Knowing His voice is essential for following Him. He is the leader. What is communicated to us by Jesus is the Word of God, His intent, and His wishes.

God's Voice
The voice carries expression, meaning, mood, and the spirit of a person. In 2 Corinthians 3:6 (NIV), the Bible says, *"...for the letter kills, but the Spirit gives life."* One thing to note is that God did not want us to just read His word and misinterpret what He said. We could not hear His voice (which spoke into nations) at that time. We also could not grasp the specific things that came from His voice and gave insight into a person's inner thoughts and feelings. Therefore, He gave us the Spirit, the Holy Ghost. Through the Spirit, we can truly understand the heart of God.

If we only read a letter or a text message, it is easy to misunderstand the writer's original intent. However, when we hear the writer's voice, the intention becomes clearer. This shows us that God's original intent was always for a close heart-to-heart, face-to-face relationship with His people. This was something that could not come from the letter. Instead, it had to be done by the Spirit. The Spirit speaks to us so that we can hear, know, and follow Jesus Christ. When we follow Him, we are able to enjoy life and have it more abundantly.

Frequency and Transmissions
Frequencies, sound waves, and electromagnetic waves are currently being used to connect people and devices for sound and other technologies (for example, wireless and cable technology). These are the mediums most often used to connect people to global networks.

Ethernet connections are used for more permanent connections. Wireless connections are used when someone or something is mobile. Our ability to hear Jesus' voice can be stationary, as well as mobile. It is stationary when we are reading or studying the Word of God. (This is the foundation of learning about God and His wisdom). When we are on the go, we can still learn from God by tapping into His voice. Hearing, knowing, and following the voice of Jesus means we must be tuned into the God frequency.

Frequency has a way of traveling. It can cover long and far distances. Wireless signals pass through us and impact our lives every day. If we had the device to receive the encrypted signal and decrypt the transmission, we would discover interactions of all sorts. For example, people are wirelessly transmitting billions of dollars each day. Those transmissions pass through us, but they do not connect to our wallets or bank accounts.

We declare that we will not miss the secure blessing that is transmitted to us. The blessing will not pass through us. Instead, we will catch it. We will receive our miracle. We will capture the thing that is released for us because we are the receivers. We can decode, interpret, and translate the spoken information (God's blessing). We receive it now! Our miracle is right here. Our bodies are responding to the frequency (of the miracle) that is designed for us. Saints, we have a secure connection for our miracle!

Beware of Interception
In the world of cybersecurity, there's a term called interception. This kind of attack allows unauthorized users to access our information, applications, or environment. Thus, targeting our privacy. Interception gives the hacker or attacker access to view and copy files, eavesdrop on our phone conversations, and read our emails. An interception attack (that is properly executed), can utilize the same frequency and duplication of our ID. This will allow hackers to intercept our information and receive what was intended for us. Once the information is retrieved, the hacker will either exploit us for money or trade our information to multiple sources on places like the dark web.

Jesus describes the interceptor in John 10:8, 10, 12-13 (KJV). There, He says, *"[8]All that ever came before me are thieves and robbers: but the sheep did not hear them...[10]The thief cometh not, but for to steal, and to kill, and to destroy: I am come that they might have life, and that they might have it more abundantly...[12]But he that is an hireling, and not the shepherd, whose own the sheep are not, seeth the wolf coming, and leaveth the sheep, and fleeth: and the wolf catcheth them, and scattereth the sheep. ¹³The hireling fleeth, because he is an hireling, and careth not for the sheep"*. The thief (the hacker) comes to steal our information, kill our system (corrupt the data), and destroy our trust and credibility.

We must be aware of the threats we face in both the cyberworld and the spiritual world. This is why Jesus said that all who came before Him were thieves and robbers. They were impersonators and imposters. The sheep will not hear the voice of an impersonator. The same principle holds true in cybersecurity, where we avoid negotiations with threat actors or hackers.

Jesus talks about the hireling, one who works for the sole purpose of money or material reward. We must be careful about the connections we make in life because some people are only there to drain us of our resources. An interceptor or impersonator does not have our best interest at heart. However, Jesus does, because He cares for His sheep. He cares about us and our eternal destiny. Today, if we hear His voice, let us submit our lives to Him and not turn away. Be encouraged to have a secure relationship with Jesus Christ, the Good Shepherd.

God's Plan for Salvation
God's plan for salvation is through Jesus Christ, who is Lord and Savior. Let us take the time to read Acts 2:38 and Romans 10:8-13. Let us believe and confess with our heart that Jesus Christ is our Lord and Savior, and we shall be saved. In doing so, we must find a Bible believing church, a body of believers that we can fellowship and grow with. Let us thrive in our relationship with the Lord. Let us receive the precious Holy Spirit that is promised to those that believe. May our secure connection to God, through Jesus Christ, bring us great joy, blessings, and most of all, abundant life.

Let us reflect on the blessings of our secure connection to God, through Jesus Christ. By doing so, we can truly appreciate the profound parallels between the divine protection of our spiritual lives and the encryption that safeguards our digital world. Just as encryption strengthens our digital communications, our relationship with God offers a sanctuary of intimacy and security that is shielded from the prying eyes of spiritual adversaries.

We have explored the dangers posed by spiritual "interceptors"—forces that seek to disrupt or steal our connection with God (much like hackers targeting unsecured networks). By understanding and fortifying this divine connection, we deepen our trust in God's unwavering protection. As a result, let us embrace the peace and confidence that not only comes from knowing our communication with the Creator is secure. It also ensures that our connection with Him will remain unbreakable.

Inspirational Downloads

-4-
ESSENTIAL UPGRADES
(Strengthening Resilience for the Future)

Upgrades

Imagine you've been relying on the same trusted software for years. It's reliable, familiar, and it gets the job done. When you hear about a new technology that promises faster results and better security, you hesitate. Why fix what isn't broken? You ignore the notifications and reminders to upgrade because you're just too busy to bother. However, before you know it, your competitors start using the new system. Suddenly, they're outpacing you, closing deals faster, protecting their data better, and adapting to market changes more efficiently. Realizing you're being left behind, you finally make the switch.

At first, there's a learning curve, but soon, you see the benefits. The new technology streamlines your work, enhances your security, and opens up possibilities you didn't even know existed. The fact is, clinging to old, familiar methods can hold us back when there's something better available. Upgrading isn't just about replacing what's old—It's about embracing what's new, in order to reach greater heights. Just as our devices need regular updates to stay secure and efficient, our lives need constant renewal. God invites us to upgrade our hearts and minds. This will ensure we're equipped to handle life's challenges with strength and resilience.

Essential Upgrades

In the journey of faith, essential upgrades are similar to vital tools that strengthen our resilience and prepare us for the challenges ahead. Just as technology requires regular updates for optimal

Essential Upgrades

performance, our spiritual lives demand continuous growth and transformation.

This chapter explores the importance of upgrading our connection with God. These upgrades, also known as Consecration, Recognition, Transformation, and Implementation (CRTI), are essential for manifesting God's purpose in our lives. As we explore these principles, we will discover how they equip us to navigate life's complexities with faith, resilience, and unwavering trust in God's providence.

What Does It Mean to Upgrade?

In computing and consumer electronics, the term *upgrade* can be defined as the process of replacing an existing application, firmware, hardware, or software with a newer or better version. The purpose is to bring the system up-to-date and/or improve its characteristics.

In today's digital landscape, cybersecurity upgrades are crucial for safeguarding sensitive information and maintaining the integrity of systems. The rapid pace of technological advancements means that cyber threats are constantly evolving. Hackers and cybercriminals are perpetually developing new methods to exploit vulnerabilities in software and hardware. Regular upgrades ensure that security systems are equipped to handle the latest threats. For example, by providing patches for newly discovered vulnerabilities and enhancing the overall resilience of the network. Upgrades in cybersecurity are also essential for compliance with regulatory standards and industry best practices. Organizations across various sectors are required to adhere to stringent security regulations, in order to protect data and privacy. Failure to comply (with regulations) can result in severe penalties and damage to the company's reputation. Executing up-to-date security measures (in a timely fashion) demonstrates a commitment to protecting consumer data and maintaining trust with clients and stakeholders. It also provides a competitive edge by showcasing a proactive approach to security.

In addition to protecting data and trust, cybersecurity upgrades help to optimize the performance of security tools and systems. As technology evolves, new security solutions become more efficient and effective. Implementing the latest upgrades can improve the detection and mitigation of threats, reduce false positives, and streamline security operations. This not only enhances protection, but also allows Information Technology (IT) teams to allocate resources more efficiently and focus on strategic initiatives. This method can prevent the teams from being bogged down by outdated and inefficient security measures.

Lastly, regular upgrades can support the overall strategic goals of an organization. Cybersecurity is not just about defense. It's also about enabling growth and innovation. By ensuring that security measures are up-to-date, organizations can confidently explore new technologies (i.e. cloud computing and Internet of Things or IoT), without exposing themselves to undue risk. Upgrades ensure that the security infrastructure is robust enough to support business expansion and new technologies, while driving growth and maintaining a strong security posture.

Our Connection to God

Just as upgrades improve and transform various aspects of the digital world, our spiritual lives can also undergo a process of transformation and improvement. In the Bible, the Apostle Paul talks about a profound spiritual transformation in his letter to the

Corinthians. This transformation can be seen as a spiritual upgrade. Let's read 2 Corinthians 3:17-18 (KJV), "*[17] Now the Lord is that Spirit: and where the Spirit of the Lord is, there is liberty. [18] But we all, with open face beholding as in a glass the glory of the Lord, are changed into the same image from glory to glory, even as by the Spirit of the Lord.*"

In these verses, Paul explains how believers, through the Spirit, undergo a continual process of transformation. This transformation is not just a superficial change but a profound enhancement of our spiritual nature, aligning us more closely with the image of Christ. Just as an upgrade is a newer version of an existing product, the Holy Spirit works within us to upgrade our spiritual lives and make us more like Christ.

In this age of technology and innovation, we are constantly faced with the need to adjust and perform upgrades to our phones, tablets, smartwatches, computers, TVs, and other electronic devices. This necessity arises from the imminent threat of vulnerabilities found in the devices. Even the networks (the data service providers for our electronic gadgets) can be affected by malicious adversaries attempting to disrupt service. In some cases, information is either stolen or held for ransom when an attacker has bypassed the defenses of a computer or network. Similarly, we seek spiritual upgrades to protect and strengthen our faith against attacks from our adversaries.

Prophetic insight reveals the enemy's strategic plan to weaken the defenses of believers and make those, who have not upgraded themselves in the Spirit, vulnerable to the wicked one's intentions. This underscores the urgent need for us to improve ourselves spiritually. It is time to move from the glory we are familiar with to a higher glory that will enable us to endure and stand resilient to the enemy's attacks.

What is Roaming?

If we take the time to understand mobile networking, roaming is a service that is available. However, we can't access it due to limitations of the service providers or unpaid services. What does this mean? Some of us have been unavailable to God. We have been in a roaming state for too long and it's time for a divine re-connection to the LORD.

If you have been roaming, this means that your current provider's limitations have been resulting in inadequate service. For example, dropped calls as you moved away from the cell tower (God); a struggle to access the internet (communion and fellowship); streams on Facebook, YouTube, and Instagram keep buffering (distractions); or you're temporarily disconnected due to unpaid services (lack of giving). We need a spiritual upgrade!

What is God telling us as He calls for our upgrade? We are undergoing a transformation to a higher level of glory. We will no longer be the same. We are now operating at a higher frequency and becoming more attuned to the voice of God. We will not rely solely on 5G, 6G, or any current wireless data network. Instead, we will connect to the ultimate source, GOD - ALMIGHTY!

What is a Hotspot?

Our positive connections with others are imperative. For example, when our cellular service is inadequate, then being in the vicinity of someone (with a better service provider) allows us to connect to that individual's hotspot. The hotspot allows us to access the same quality of service the other individual has. Remember, no matter how faithful we are to our service provider, we will only receive what that provider can offer! Similarly, wireless networks can be compared to the Christ-centered network of Apostles, Prophets, Pastors, Teachers, and Evangelists given to the Church. Just as we rely on strong wireless connections (to stay linked to the digital world), by connecting to a ministry gift, we can tap into the divine service (Grace and Glory) provided to us.

A prime example of having a positive connection with others is when Elisha tapped into the service of Elijah. Elisha was so close to Elijah that the sons of the prophets recognized the spirit of Elijah resting on him (2 Kings 2:15). Because Elisha not only submitted to Elijah, but also tapped into his service, Elisha received a double portion of Elijah's spirit. This demonstrates that submission is essential for impartation.

Had Saul submitted himself to God, he would have walked among the prophets. In 1 Samuel 19, Saul sent messengers to capture David. The Bible recounts that when the messengers saw the company of prophets prophesying (with Samuel leading them), the Spirit of God came upon the messengers, and they also prophesied. This happened three times and each time, the messengers prophesied. Even when Saul pursued David himself, the Spirit of God came upon him, and he prophesied from Sechu to Naioth in Ramah. The Spirit of God allowed Saul and his messengers to tap into the prophetic frequency of that location. When we recognize the company we are in, we can tap into the grace that is on their lives. May the Spirit of the LORD God move upon us, now, to tap in!

Our Connection to God for Manifestation

As we continue to seek spiritual growth and elevation, it is crucial to understand the importance of alignment and connection with those who carry divine grace and wisdom. We are being called to arise, emerge, and manifest for a purpose. Our manifestation of who we are becoming is closely tied to whom we are connected and submitted to. For example, Samuel was connected to Eli. In 1 Samuel 3:1, we discover that Samuel ministered to the LORD before Eli, meaning he was assisting Eli or operating under his guidance. When the lamp or candlestick in the temple went out and God called Samuel, Samuel thought it was Eli, who called him. (Read 1 Samuel 3:1-10).

Spiritual leaders are very important in our lives. Eli was a spiritual father to Samuel and gave him the decryption code to connect to God for himself. Oftentimes, we need the decryption code to make a proper connection to the right network (Wi-Fi). We also need the Multifactor Authentication Code (MFA) to gain full access to an application (i.e. Office365). The code Eli gave to Samuel was simple: "Go, lie down; and it shall be, if He calls you, that you must say, *Speak, Lord, for your servant hears*" (1 Samuel 3:9 KJV).

God uses a voice that we can understand and listen to. We need someone who can speak the Word of the LORD into our lives for deliverance and preservation. Hosea 12:13 (KJV) says, "And by a prophet the Lord brought Israel out of Egypt, and by a prophet was he preserved." Who is the prophet in your life?

In 1 Samuel 3:1, the Bible tells us that the word of the LORD was precious in those days and there was no open vision. There was no open vision for the seer prophet until Eli gave Samuel the password (key or access) to the open door. Having access allowed Samuel to be elevated to hear the voice of God and experience the open vision. (1 Samuel 3:11-15 KJV). In verse 15, the Bible says, "And Samuel lay until the morning and opened the doors of the house of the LORD. And Samuel feared to show Eli the vision."

Some of us will experience elevation. It will come through our commitment to having and maintaining the right connection in our lives. Below are four points of wisdom that will empower us for our time of manifestation.

1. **Our connection for manifestation requires _consecration_.** Consecration means being set apart for God's service. In 1 Samuel 3:1, we see that Samuel ministered to the LORD before Eli. Our spiritual identity is recognized by who we are connected to. If we are connected to our spiritual leaders, their influence will be evident in us.

   Today, the church faces a growing concern of identity crisis. The enemy's plan is to undermine godly character and create division between fathers and children (Malachi 4:5-6). Let us pray and declare that God will reconcile the hearts of fathers to their children and children to their fathers. Godly character and consecration will once again be evident in the lives of God's people. The sons of God will emerge and be manifested!

2. **Our connection for manifestation requires _recognition_.** We must reach a point in our lives where we accept the truth that with God, all things are possible. It is essential to acknowledge God in everything we do. Proverbs 3:5-6 encourages us to, *"⁵Trust in the Lord with all your heart; and lean not on your own understanding. ⁶In all your ways*

acknowledge him, and he shall direct your paths." To acknowledge him is to recognize him. The scripture makes it clear that if we recognize the LORD in all our ways, he will guide us in our elevation by making our paths straight and directing us in the way we should go.

3. **Our connection for manifestation requires _transformation_.** We must be adaptable and open to change. Romans 12:2 says, *"And be not conformed to this world: but be ye transformed by the renewing of your mind, that ye may prove what is that good, and acceptable, and perfect, will of God."* God will speak words of elevation to us that necessitate a complete inner transformation. We are in a time and season where transformation is crucial for where God is leading us. The word God speaks in this season is transformative, elevating the Body of Christ to operate above the frequencies of chaos and calamity. When our transformation takes place, our minds will be renewed. Our spirits will be revived. We will be released into effective ministries and rewarded for our faithfulness!

4. **Our connection for manifestation requires _implementation_.** We must actively engage and carry out the plans and courses of action God requires. Let us execute the Plan! Let us be inspired by Daniel 11:32, *"...But the people who know their God shall be strong, and carry out great exploits."* Like Samuel, we must hear and respond to God's voice. Our response involves not only acknowledging His words but also faithfully carrying out His commands. This requires us to walk by faith, not by sight.

In navigating life's journey, essential upgrades are crucial for strengthening resilience and preparing for future challenges. Just as technology requires regular updates to maintain optimal performance, our spiritual lives also benefit from upgrades. Our connection with God and spiritual leaders plays a pivotal role in this process, guiding us towards greater spiritual maturity and effectiveness.

My prayer is that the voice of God would strengthen our faith. May our elevation come as we experience God's promotion through submission to Him, and the spiritual authority in our lives. May the LORD open the door to His storehouse of wisdom and understanding. May we have the grace to walk through that door and enter into a place of lasting transformation. Let's connect, elevate, and experience the true power of the LIVING GOD!

In conclusion, our connection for manifestation requires: *Consecration, Recognition, Transformation, and Implementation (CRTI)*. Through consecration, we are set apart for God's purposes; Through recognition, we acknowledge His guidance in all aspects of life; Through transformation, we embrace inner renewal for alignment with His will; and Through implementation, we actively carry out His plans.

As we upgrade spiritually, we will enhance our resilience and readiness to face whatever the future holds as long as we are empowered by God's wisdom and grace.

Inspirational Downloads

-5-
THE MAN IN THE MIDDLE
(Interception or Intercession)

> **Man-in-the-Middle**
> Imagine you are emailing a sensitive document to a colleague. You hit send and you feel confident the email is on its way. However, what you don't know is that, while in transit, the email was intercepted by hackers. They altered it, injected a malicious code, and made it look like it was still from you. When your colleague received the email, everything seemed normal, but the email had been compromised. This scenario highlights how our unprotected communications can be vulnerable to interference. Encryption secures our data and protects it from the malicious "Man in the Middle".

Man-in-the-Middle (MitM) can be defined as an attack, in which an attacker is positioned between two communicating parties to intercept and/or alter data traveling between them. MitM cyber-attacks pose a serious threat to online security because they give the attacker the ability to capture and manipulate sensitive, personal information (i.e. login credentials, account details, or credit card numbers in real time).

In this chapter, we'll explore a profound contrast. Although technology can be exploited, there exists a divine "Man in the Middle", who intercedes on our behalf. His name is Jesus, the ultimate intercessor. He doesn't have malicious intent. Instead, He stands between us and God, ensuring that our prayers are perfectly conveyed and safeguarded. As our intercessor, Jesus preserves the integrity of our communication with God and He protects it from any distortion or interference. Through Him, our words reach the Father exactly as intended, pure, and unhindered. Discover how He stands between us and the forces that seek to harm us. Explore the power of His love, protection, and grace as we dive into the captivating connection between cybersecurity and our spiritual journey.

During MitM attacks, cybercriminals insert themselves in the middle of data transactions or online communication. A very common type of MitM attack is focused on web browser infection and the injection of malicious proxy malware into a victim's device. The malware is commonly introduced through phishing emails. As a result, the attacker gains easy access to the user's web browser and the data it sends and receives during transactions. Online banking and e-commerce sites (which require secure authentication) are the prime targets of MitM attacks. This is because they enable attackers to capture login credentials and other confidential information.

There are several ways to prevent MitM attacks. For example, by using Secure Connections, Virtual Private Network Encryption (VPN), and Endpoint Security. Below are additional methods to help users prevent MitM attacks:

1. Only visit websites that show "HTTPS" in the URL bar, instead of just "HTTP". Most browsers display a padlock sign before the URL, which indicates a secure website.

2. Avoid using unsecured public Wi-Fi connections, as they are susceptible to attacks and interception by cybercriminals.

3. Use multifactor authentication on online accounts. It adds an additional layer of security to online communications.

4. Avoid phishing emails.

5. Use a VPN that encrypts internet connections and online data transfers (e.g. passwords and credit card information). It should be used when connecting to unsecure public Wi-Fi networks and hotspots.

6. Have anti-malware and internet security products in place on all internet capable devices.

A MitM attack shows how vital it is for you to protect sensitive information from being intercepted. Just as a Man-in-the-Middle intercepts and mediates communication between two parties, Jesus is the mediator and intercessor between God and mankind. Jesus bridges the gap between humanity and God. He is the advocate, interceding on our behalf. Jesus is the bridge-builder in the relationship between God and His people. He offers forgiveness, guidance, and salvation. Let's Discover more about Christ Jesus the Mediator.

Christ Jesus the Mediator
A mediator is often seen as one who intervenes between two parties to make or restore peace and friendship, to form a compact, or to ratify a covenant. We find Christ Jesus the mediator in 1 Timothy 2:5 (KJV). It reads, *"For there is one God, and one mediator between God and men, the man Christ Jesus"*.

The book of Hebrews 4:14-15 speaks of Christ Jesus as a great high priest. It reads, *"¹⁴Seeing then that we have a great high priest, that is passed into the heavens, Jesus the Son of God, let us hold fast our profession. ¹⁵For we have not a high priest which cannot be touched with the feeling of our infirmities; but was in all points tempted like as we are, yet without sin".*

Having read the preceding text, it is important to recognize that Jesus comprehends your vulnerabilities. He relates to your imperfections. He underwent temptations in every form, much like the temptations you encounter. Yet, Jesus remained sinless, never succumbing to the enticements that crossed His path. This realization should elicit a sense of enthusiasm within you. When you approach God in prayer, you can do so with unwavering assurance, knowing that you will obtain mercy and discover grace to assist you in achieving your breakthrough.

Another valuable truth of how Jesus (as the intercessor) impacts your life is found in Hebrews 7:25 (NKJV). The scripture reveals, *"Therefore He is also able to save to the uttermost those who come to God through Him, since He always lives to make intercession for them"*. Through your connection with Jesus, you can experience complete and eternal salvation before God. This embodies the profound beauty of salvation through Jesus Christ. His living presence allows for your comprehensive and everlasting redemption. As the living Savior, Jesus fulfills the role of a divine intermediary, who stands as the bridge between God and humanity—the very essence of the Man-in-the-Middle. Praise and glory be to God for this magnificent reality!

The Power of Prayer
Being an intercessor, prayer formed the core of Jesus Christ's relationship with the Father. His prayers encompassed his own needs, his mission, and his disciples. This practice endures as he

intercedes on behalf of all believers. As a faithful high priest, Jesus consistently intercedes for those who embrace his name. The act of prayer remains ceaseless, serving as a reminder of the necessity for daily and unwavering communion with God. This truth is echoed in Luke 18:1, which asserts that individuals are obligated to maintain an unbroken practice of prayer, refusing to succumb to weariness.

Prayer provides you with strength during challenges and bestows you with the capacity to endure hardships. It shapes the caliber of your triumphs in Christ. When you neglect prayer, there is a risk of spiritual depletion that can leave you vulnerable to the adversary—the figurative "man-in-the-middle." This opponent, often personified as the devil, seeks to assail, intercept, and bring an end to your spiritual well-being. John 10:10 (TLB) says *"The thief's purpose is to steal, kill and destroy. My purpose is to give life in all its fullness"*.

Being aware of the active adversary, who opposes God's plan for your life, it becomes imperative to engage in prayer. Jesus thoughtfully instructed his disciples in the art of prayer, recognizing that their effectiveness in their divine calling depended on their ability to communicate with God. His disciples observed his dedication to prayer. They also noticed the considerable time he invested in this practice. Furthermore, they witnessed the tangible outcomes of Jesus' prayers throughout his earthly ministry. Jesus understood that prayer paved the way for the manifestation of divine power in one's life. This very power and authority empowered him to perform miraculous healings, raise the dead, restore sight to the blind, cast out demons, and bring about restoration to the afflicted. Jesus grasped that this authority could only be accessed through a direct relationship with God, fostered by devoted prayer and communion in His divine presence.

The Prayer of Intercession

In addition to devoting time to prayer and instructing his disciples on prayer, Jesus also engaged in intercession on behalf of his disciples, extending this intercessory role to include you as well. You will explore four facets of Jesus' intercession within the passage of John 17:11-17.

UNITY is the state or quality of being one. In John 17:11 (TLB), Jesus' initial entreaty to the Father pertained to unity. The scripture states, *"Now I am leaving the world, and leaving them behind, and coming to you. Holy Father, keep them in your own care—all those you have given me—so that they will be united just as we are, with none missing"*. In His plea, Jesus implored the Father to safeguard His disciples, so they would stand united, mirroring the unity between Him and the Father. This prayer held immense importance considering the challenges and shared experiences awaiting the disciples during their ministry. The effectiveness of Jesus' instructions to his disciples hinged on their unity. Collaborative effort and synchronized thinking were imperative. This underscores the necessity of unity in Christ. It is a direct result of shared purpose. Purpose acts as both the driving force and motivation behind fulfilling God's commands. This unifying force is devoid of self-interest, as individuals might attempt to unite for concealed agendas.

Frequently, an agreement (driven by self-interest) is accompanied by a concealed motive aimed at causing division or prioritizing one party over another. The unity that Jesus fervently prayed for was rooted in a direct plea to God, contrasting with a self-centered approach. Jesus' actions consistently revolved around aligning with the Father's will. He regularly sought divine guidance. Therefore, in John 17:11 (NIV), Jesus requested, *"...Holy Father, protect them by the power of your name, the name you gave me, so that they may be one as we are one"*. In this context, protect means to "safeguard", which implies the acts of preservation, possession, and salvation. The Father, being the ultimate protector, harbors the desire to preserve and save you. Moreover, He maintains the role of Creator, shaping you much like a potter molds clay. His craftsmanship molds you into an integral component of a cohesive community of Believers. This community is infused with the fullness of the Holy Ghost and imbued with divine power.

When the Church, comprising the community of Believers in Christ, unites, we align ourselves with God's purpose, His authoritative Word, and His divine directive. Amos 3:3 (NIV) poses a thought-provoking question, which asks, *"Do two walk together unless they have agreed to do so?"* The Church's collective agreement propels us as an indomitable force, advancing the Kingdom of God without impediment. The potency of Unity bestows upon us a remarkable DIVINE ADVANTAGE! Psalm 133:1 (ESV) joyously affirms, *"Behold, how good and pleasant it is when brothers dwell in unity"*.

The unity depicted in the Bible is rooted in spiritual vitality and adherence to the Word of God. Those genuinely saved experience spiritual companionship, recognizing each other and disassociating from the forces of darkness. Ephesians 5:11 (NIV) underscores this, advising, *"Have nothing to do with the fruitless deeds of darkness, but rather expose them"*. Philippians 1:27 (NIV) directs us to *"...stand firm in the one spirit"*. Nowhere in the New

Testament is division among God's people endorsed. Conversely, scripture consistently champions unity while condemning division. Throughout the Bible, the emphasis resounds on the singularity of Lord, salvation, God, faith, Spirit, mind, mouth, eye, body, baptism, new and living way, and Savior. To wholeheartedly serve the true and living God, you must follow the one Lord through the new and living path He sanctified for those who heed His guidance. Embrace His revealed and indwelling Word, being *unified in Christ Jesus*.

JOY gives us a feeling of great pleasure and happiness caused by something exceptionally good or satisfying. The subsequent appeal (Jesus made to the Father) pertained to the bestowal of JOY in John 17:13 (KJV). It reads, *"And now come I to thee; and these things I speak in the world, that they might have my joy fulfilled in themselves"*.

Jesus offered a prayer for the disciples, seeking to fill them with joy. Recognizing that the path of ministry would present challenges, Jesus requested that his disciples experience the same boundless joy that resided within him. This joy brings forth a heightened sense of enthusiasm and endurance. Nehemiah 8:10 (KJV) declares that *"...the joy of the Lord is your strength"*. It is imperative to forge ahead in life, regardless of the circumstances you encounter. How is this achieved? It is achieved by embracing the joy of the LORD, which empowers you to persevere. This joy also instills a sense of rejoicing in the LORD. Habakkuk 3:18-19 (KJV) illustrates this sentiment, *"[18]Yet I will rejoice in the Lord, I will joy in the God of my salvation. [19]The Lord God is my strength, and he will make my feet like hinds' feet, and he will make me to walk upon mine high places"*. Beloved, find joy in the God of your salvation. Embrace the joy that strengthens you to rise above challenges and walk confidently in elevated circumstances.

Reasons for rejoicing are always present. Reflect on the comforts that grace and mercy from God have bestowed upon you in life. Your education, wealth, health benefits, and the doors divinely opened are blessings to be deeply grateful for and to find joy in receiving. I affirm that even amidst life's storms, you possess the capacity to experience joy. Storms are crucial experiences that provide you the chance to navigate and triumph over them, all while relying on the joy of the LORD throughout the journey! You will successfully navigate trials and tests. Through Jesus Christ, your Lord and Savior, you emerge as a conqueror.

In John 17:15-16 (KJV), Jesus presented a third entreaty to the Father. It was centered on seeking **PROTECTION**. The scripture reads, *"[15]I pray not that thou shouldest take them out of the world, but that thou shouldest keep them from the evil. [16]They are not of the world, even as I am not of the world"*. Protect means to keep one safe from harm or injury; to defend or guard from attack, invasion, loss, annoyance, or insult.

Jesus prayed that God would protect his disciples from the attacks of the Evil One. It is certain that in this world there is a power of evil which is in opposition to the power of God. It is inspiring to know that God is watching over your life to guard you from the assaults of the enemy. *"God is our refuge and strength, a very present help in trouble"* (Psalm 46:1 KJV). When the enemy uses people to plot against you and come up with devious schemes to cast you down, their plans will FAIL! What the LORD has ordered for your life cannot be stopped. Once the LORD has unlocked a door for you, no force can seal it shut. When God has bestowed His blessings upon you, no man can speak a curse against you. The LORD made this promise in Isaiah 59:19 (KJV), *"...When the enemy shall come in like a flood, the Spirit of the LORD shall lift up a standard against him"*. When the attacker pursues you, our Advocate will block the trajectory of the enemy's attack!

In truth, as a Believer in Christ, you have been given power over your enemy. Luke 10:19 (KJV) declares, *"Behold, I give unto you power to tread on serpents and scorpions, and over all the power of the enemy: and nothing shall by any means hurt you"*. Once you have been granted divine power, fear no longer has a place in your life. Your existence is governed by the Word of God and unwavering faith in His declarations. God's affirmation, *"No evil shall befall you"* (Psalm 91:10 NKJV), holds true. Every assault from the adversary is incapable of breaching the protective fortifications that the LORD has encircled around your being. The charge to safeguard and watch over you in all your endeavors has been entrusted by the LORD to His angels. You have Divine protection!

The scripture passage found in Luke 10:19 (NIV) proclaims that through the reception of divine power, you shall possess the authority to conquer serpents and scorpions, prevailing over all the might of the adversary. Operate with the strength of God's Spirit, allowing for the manifestation of signs, wonders, and miracles that will be readily apparent to numerous witnesses. Let the deeds of the devil be dismantled in the process.

No matter what conflict or challenge you are currently grappling with or might encounter in the days ahead, the word of the LORD speaks directly to this situation. Psalm 121:7-8 (NKJV) proclaims, *"⁷The Lord shall preserve you from all evil; He shall preserve your soul. ⁸The Lord shall preserve your going out and your coming in from this time forth, and even forevermore"*. The LORD stands as your guardian. Now is a good moment to offer praise to Him and shout, *"Thank you, Jesus!"*

In John 17:17 (KJV), the fourth and last request Jesus presented to the Father was for **SANCTIFICATION**. The scripture declares, *"Sanctify them through thy truth: thy word is truth". Sanctify can be defined as making holy or setting apart as sacred; consecrate; to purify or free from sin.*

This aspect of Jesus' prayer is transformative. Its transformative nature lies in its ability to guide believers of Christ toward a shift in both their behavior and speech, fostering a change in their way of life. The Word of God not only bestows joy and tranquility upon you, it also instills purity within your being. Sanctification propels you towards a deeper connection with God through Jesus Christ and empowers your journey towards a life of holiness. Strive diligently to lead a life devoted to God. Hebrews 12:14 (NIV), declares, *"Make every effort to live in peace with everyone and to be holy; without holiness no one will see the Lord"*. My friend, holiness represents a way of living that incorporates both God and humanity.

Your pursuit after God requires being obedient to His Word. The Bible teaches that the Word of God is TRUTH. Therefore, it is imperative that you apply God's word to all aspects of life. God's word operates in such a way that it precisely accomplishes what is declared. Given that you have embraced these invaluable promises from the Word of God, 2 Corinthians 7:1 (NKJV) directly addresses you by stating, *"Therefore, having these promises, beloved, let us cleanse ourselves from all filthiness of the flesh and spirit, perfecting holiness in the fear of God"*. Beloved, your consecration (*being set apart for God's purpose*) is achieved through the ministry of God's word. John 15:3 (KJV) proclaims *"Now ye are clean through the word which I have spoken unto you"*. The spoken word, the Rhema, actively purifies and consecrates your existence. As you embrace God's Word, you will find rejuvenation, renewal, sanctification, and edification!

In exploring Christ Jesus as the Mediator, the Power of Prayer, and His Prayer of Intercession, you have discovered that it is essential for believers to uphold their commitment to prayer and unwavering faith in the Word of God. Despite your busy schedule, consistently allocate time for prayer, examine the Scriptures, and integrate their teachings into your daily routines. Believers in Christ, who neglect prayer, find themselves devoid of power. This neglect of prayer is among the factors contributing to a life of defeat for some. Beloved, persist in your prayerful connection and alignment with God's purpose for your life. Remember the words of Luke 18:1, which emphasize the necessity of unceasing prayer, which will prevent weariness and discouragement.

In conclusion, this chapter examined the contrast between the man-in-the-middle attack (interception) and Jesus Christ, who mediates for the greater good (intercession). With interception we saw how a malicious attacker intervenes and alters the trajectory of data in transit. With intercession we saw the profound impact of Jesus Christ as The Man in the Middle, who intercedes on our behalf. Our relationship with Jesus shapes our outcomes and ultimately fosters a healthy relationship with Him and others.

Inspirational Downloads

-6-
EMBRACING CHANGE
(The Power of Rebooting Your Life)

> **Reboot**
> Imagine running your computer for days, even weeks, without a break. You notice some significant changes. For example, programs start to lag, the screen freezes, and everything feels sluggish. You keep pushing through, ignoring the signs, until finally, nothing works as it should. Frustrated, you finally acknowledge that it is time for a reboot. You shut the computer down, let it rest for a moment, and then power it back on. Instantly, everything runs smoother and faster, like it's brand new. This scenario is a simple reminder that sometimes, we just need to reboot. When we push ourselves too hard without rest, we become drained, unfocused, and less effective. Just as our computers need a reboot to refresh and function at their best, we need moments of rest and renewal in our lives. God invites us to pause, take a break, and allow Him to refresh our spirits. This will allow us to move forward with clarity, strength, and purpose.

Life, much like a complex computer system, often encounters moments of stagnation or unforeseen glitches. In these instances, we may find ourselves yearning for a fresh start or a complete reboot. This chapter examines the intricacies of a transformative reboot in the context of our lives. We explore the power of embracing change, the significance of learning from our past, and the transformative journey that unfolds when we choose to reboot. Let's embark on this thought-provoking exploration of rejuvenation and growth. It will help us to understand how embracing a reboot can be the catalyst for remarkable transformations in our lives.

Embracing Change Page 74

The concept of rebooting is integral to the world of information security. There, it signifies the protection and renewal of digital systems. With the increasing reliance on technology for communication, commerce, and data storage, safeguarding sensitive information has become more important than ever. Information security practices (i.e. implementing strong passwords, encrypting data, employing firewalls, performing regular software updates, and using antivirus programs), ensure the confidentiality, integrity, and availability of information. This chapter will draw parallels between these digital safeguards and the steps we can take to reboot our lives. It will also highlight how such measures are essential for maintaining trust, reliability, and overall well-being in digital and personal spaces.

Rebooting (an often overlooked, but vital IT term), refers to the process of restarting a computer system. This action can be as simple as turning a device off and then on again. Its benefits are substantial. Rebooting helps to clear the system's memory, shut down malfunctioning processes, and apply updates or configuration changes. It is a common troubleshooting step that resolves a myriad of technical issues, ranging from slow performance to software glitches. By rebooting a system, users can refresh the operating environment and ensure that applications run smoothly and efficiently. Regular reboots can prevent minor issues from escalating into major problems, thereby maintaining the stability and performance of the system.

In the context of information security, rebooting plays a crucial role. It can help in the application of security patches and updates that protect systems from vulnerabilities and cyber threats. When a system is rebooted, it ensures that newly installed updates take effect and potential security loopholes are closed. Furthermore, rebooting can disrupt ongoing cyber-attacks by stopping malicious processes and preventing attacks from occurring. For example, those attacks which involve malware or ransomware. Regularly scheduled reboots can thus be an integral part of a robust security strategy. It can also be conducive to maintaining the integrity and safety of critical information systems.

To maximize the benefits of rebooting (for personal computers, home networks, or larger organizational systems), it is essential to follow best practices. Users should save their work and close all applications before initiating a reboot. (This will prevent data loss). For home networks and personal use, it's beneficial to reboot devices regularly to maintain optimal performance. Scheduling reboots during maintenance windows can minimize disruption for both home and organizational settings. For servers and critical systems, automated scripts and management tools can be utilized to ensure reboots occur without manual intervention. Additionally, organizations should educate their employees on the importance of rebooting. They should also incorporate it into their IT policies. By treating rebooting as a routine maintenance task (rather than a last resort), individuals and organizations can enhance the overall health and security of their IT infrastructure.

Spiritually, the themes of renewal, transformation, and new beginnings align perfectly with the concept of rebooting. Sometimes in life, we discover that we need a reboot to take place. When we go through traumatic experiences, we may find the need to renew our minds. In Romans 12:2 (KJV), the Bible says, *"...be ye transformed by the renewing of your mind..."* Renewing is a synonym for rebooting or restarting. When we experience an interruption in our lives, we must begin again or return to where we left off. When growth and progress have been hindered for too long (based on our way of thinking), a mental restart may be necessary for us to move forward.

After the children of Israel experienced the loss of Moses, they mourned for thirty days. When the time of mourning and weeping was over, the LORD spoke to Joshua. God made an important statement to prompt a mental reboot for Joshua and the children of Israel. Three times, God said, *"Be strong and very courageous!"* (Read Joshua 1:6, 7, 9). This directive was crucial for their renewal and transformation.

God speaks to bring us to a place of obedience and perfection. Although Jesus was the Son of God, He learned obedience through suffering (Hebrews 5:8). Hebrews 5:9 tells us that once Jesus achieved perfection, He became the source of eternal salvation for all, who obey Him. Perfection is tied to our obedience to God. Our will is to do His will. Joshua was positioned to walk in this obedience to see God's perfect work accomplished through his life. To do this, Joshua had to be strong and courageous.

To be strong and courageous requires a certain mindset. Strength is not only seen in our physical construct, but also experienced in our mental capacity. To deal with conflicts, unforeseen circumstances, loss, or rejection, we need to be strong mentally, emotionally, and spiritually. For this to happen in our lives, we must be strong in our faith. Faith is the catalyst for a reboot. It empowers us to stand on the promises of God. In Romans 4:19-20 (KJV), the Bible illustrates the power of faith in Abraham's life. The passage reads, *"[19]And being not weak in faith, he considered not his own body now dead, when he was about an hundred years old, neither yet the deadness of Sarah's womb. [20]He staggered not at the promise of God through unbelief; but was strong in faith, giving glory to God"*.

To produce a son, Sarah needed a reboot of her reproductive system. Through faith, it became possible, and she bore Isaac. Our time to produce is not over. God will revive us and allow us to experience a reboot so our lives can be productive. We shall accomplish what the LORD has declared over our lives. We will give birth to our dreams and see them manifest.

Not only must we be strong, but we must also be courageous, or *"of good courage"*, as God told Joshua. In the face of fear, we must be brave and not intimidated by what we experience. Therefore, courage is necessary. As we hear the Word of the LORD, strength and courage are released into our lives. We will experience a REBOOT, changing our perspective on the situation. We are released from fear and intimidation. Our sadness is turning to joy. Life might be difficult at times, but the LORD empowers us to move on. Our situations will not defeat us. We are equipped (by God) to handle what we face and where He is leading us. This truth is echoed in the Bible.

Isaiah 43:18-19 (NIV) says, *"[18]Forget the former things; do not dwell on the past. [19]See, I am doing a new thing! Now it springs up; do you not perceive it? I am making a way in the wilderness and streams in the wasteland"*. This passage speaks to us. It reveals a new version of ourselves and encourages us to walk in the new because we are new creations. *"...old things are passed away; behold, all things are become new"* (2 Corinthians 5:17, KJV). This means stepping out of our old mindsets and embracing a transformed way of thinking. In difficult moments, the Lord promises to make a way. He said, *"...I will do a new thing"* (Isaiah 43:19 KJV). New represents potential and opportunity. It manifests change and things that are unfamiliar. God is the originator of new things. Because He creates every good and perfect gift, we can decree that good things are coming our way. James 1:17 (KJV) declares, *"Every good gift and every perfect gift is from above, and cometh down from the Father of lights, with whom is no variableness, neither shadow of turning"*.

GOD WILL RENEW

We declare that God, the originator of new beginnings, will renew. Just as God bestows every good and perfect gift upon us, a reboot offers us the opportunity to step back and receive our own renewal. This is similar to when Jesus withdrew into the wilderness. This withdrawal was necessary for Jesus to be updated and rebooted. As a result, He emerged with greater efficiency and readiness to do the will of God. When this reboot happens in our lives, we move faster and with more momentum, as God works for us and within us.

When we experience this reboot, we must embrace the new operating structure or the new way of functioning. We are called to be new creatures in Christ, putting away old mindsets and adopting a transformed way of thinking. The Lord is doing a new thing for us. He is making a way in the wilderness and streams in the wasteland. As we walk in this newness, we can trust that God's good and perfect gifts are coming our way. He is empowering us to move forward with renewed strength and purpose.

Experiencing the NEW happens for us as Believers in Christ. 2 Corinthians 5:17 (KJV) says, *"Therefore if any man be in Christ, he is a new creature: old things are passed away; behold, all things are become new"*. We are new people, new beings, and new creations. We are marvelous, appealing, and renewed. As a matter of truth, the Bible tells us that God will beautify the meek with salvation (Psalm 149:4 KJV). We are fearfully and wonderfully made (Psalm 139:14 KJV). The Lord said in His Word, *"[22]That ye put off concerning the former conversation the old man, which is corrupt according to the deceitful lusts; [23]And be renewed in the spirit of your mind; [24]And that ye put on the new man, which after God is created in righteousness and true holiness"* (Ephesians 4:22-24 KJV). We are renewed daily! 2 Corinthians 4:16 (KJV) says, *"For which cause we faint not; but though our outward man perish, yet the inward man is renewed day by day."* We will not faint in this season. We will not get tired. We are being renewed in the Spirit!

We also have a future hope of NEW. Our hope ultimately rests in the promise that God will make all things new. In Revelation 21:5-7 (KJV), it is written, *"[5]And he that sat upon the throne said, Behold, I make all things new. And he said unto me, Write: for these words are true and faithful. [6]And he said unto me, It is done. I am Alpha and Omega, the beginning, and the end. I will give unto him that is athirst of the fountain of the water of life freely. [7]He that overcometh shall inherit all things; and I will be his God, and he shall be my son."* We have hope if we are Overcomers. It is very important for us to overcome whatever it is that is not of Christ in this season. We cannot entangle ourselves with the spirit of this age (Laodicea) and the deceptions of our day.

GOD WILL RESTORE
We declare that God will restore. God has the ability to restore. He restores us to a right relationship with Him through forgiveness and justification. He is able to restore earthly relationships. He can even restore days and years that have been lost. Joel 2:25 (KJV) declares, *"And I will restore to you the years that the locust hath*

eaten, the cankerworm, and the caterpillar, and the palmerworm, my great army which I sent among you". God in His great mercy brings restoration. Not only can He renew a life and redeem its future, but he can also redeem its past. He restores the years of our calamity. Glory to GOD!

In Scripture, we see God's power of restoration countless times. When Jacob was finally reunited with his lost son Joseph, he described the grief-filled days of his life as "few and evil" (Genesis 47:9 KJV). However, in his last days, by God's mercy, Jacob was able to look back on his life and see that God had been his shepherd all along. Additionally, he saw that God redeemed him from the evil that once marked his life. He made a declaration over Joseph that gives us insight into his restoration.

In Genesis 48:15-16, Israel (Jacob) blessed Joseph, saying, *"[15]God, before whom my fathers Abraham and Isaac did walk, the God which fed me all my life long unto this day, [16]The Angel which redeemed me from all evil, bless the lads; and let my name be named on them, and the name of my fathers Abraham and Isaac; and let them grow into a multitude in the midst of the earth"*. The father's blessing establishes an identity, a reputation, a uniqueness, and a name that will not be denied. Whatever blessing that has been established by God, in our names, may we receive it. Blessings of favor! Blessings of expansion! Blessings of greatness! Blessings of prosperity! Let it all be upon us. Glory to God!

In the story of Ruth, God took a family whose name faced extinction and restored to them a secure future. He joined them into His plan of redemption by placing them in Jesus' lineage. Ruth was a child of the Moabites. The Moabites came from the lineage of Lot who came out of Sodom and Gomorrah with his two daughters. He had no son, no heir, and so his daughters came up with a plan to give him sons. In Genesis 19:37, the firstborn daughter of Lot bore a son and called his name Moab (the same is the father of the Moabites until this day). Ruth 1:22 (KJV) reads, *"So Naomi returned, and Ruth the Moabitess, her daughter in law, with her, which returned out of the country of Moab: and they came to Bethlehem in the beginning of the barley harvest"*. It was in the gleaning of the harvest (*Shavuot or Pentecost*) that she met Boaz, the owner of the field and the kinsman redeemer. When Boaz took Ruth to be His wife, they had a son and called his name Obed (the restorer of life). Obed is the father of Jesse, who is the father of David. From David, we can trace the genealogy to Jesus Christ, the son of David (Matthew 1:1-17). Jesus is the royal heir to the throne of David, and he will sit upon the throne to rule the earth. In that day (millennium), the LORD shall be king over all the earth and there shall be one LORD (Zechariah 14:9 NIV)!

GOD WILL RENAME
We declare that God will rename. Names carried a lot of significance in scripture. Throughout the Bible, people are introduced to us by name and names have meanings. Eve is *the mother of all the living* (Genesis 3:20 KJV). Isaac is *laughter* (Genesis 21:6 KJV) and Samuel is *God Heard* (1 Samuel 1:20 KJV). What is even more significant is the renaming of people in Scripture. When God gave someone a new name, it was always a sign of renewed purpose, and a redeemed life based on a covenant relationship.

God changed Abram's name to Abraham to signify His promise to make Abram the father of many nations (Genesis 17:5). He changed the names of Hosea's children from *No Mercy* and *Not My*

People to *My Loved One* and *My People* to symbolize his love for Israel and his plan to redeem her from idolatry (Hosea 1 KJV). When Simon and Saul became Jesus' disciples, Simon became Peter (Matthew 4:18) and Saul became Paul (Acts 13:9). They received new identities in Christ as they forsook their old life and became servants of Jesus Christ.

When we come into a relationship with Christ Jesus, we become members of the Body of Christ. 1 Corinthians 12:27 (KJV) declares, *"Now ye are the body of Christ, and members in particular"*. This body of Christ is the CHURCH (a living organism). In Revelation 3:7 (KJV), Jesus speaks to the angel of the Church of Philadelphia, saying, *"...These things saith he that is holy, he that is true, he that hath the key of David, he that openeth, and no man shutteth; and shutteth, and no man openeth"*. The one who is the THAT (in "I AM THAT I AM") says that He is holy. He is true. He has the key of David. He opens and no man shuts. He shuts and no man opens. THAT is the person, thing, idea, state, event, time, or remark indicated, mentioned, pointed out, or understood from a situation. I am He that is holy. I am He that is true. I am He that is the door.

In Revelation 3:8 (KJV), Jesus speaks of Himself and declares to the Church of brotherly Love (*Philadelphia*), *"I know thy works: behold, I have set before thee an open door, and no man can shut it: for thou hast a little strength, and hast kept my word, and hast not denied my name"*. When God opens a door for the Church, it is time to ADVANCE. The Church of Philadelphia is loved by Jesus. Jesus says in Revelation 3:12, *"Him that overcometh will I make a pillar in the temple of my God, and he shall go no more out: and I will write upon him the name of my God, and the name of the city of my God, which is new Jerusalem, which cometh down out of heaven from my God: and I will write upon him my new name"*. His new name will be written upon those that will overcome. A new name means a new identity and purpose with Jesus Christ. All the blessings that are in the new name will be upon the overcomer! Where He is, there shall we be also!

GOD WILL REVIVE

We declare that God will REVIVE. God makes dead things alive again. He has power over death in every sense and He demonstrated that to us when He raised Jesus from the dead. The scripture says that, as believers, we have that same power dwelling in us. Romans 8:11 (KJV) says, *"But if the Spirit of him that raised up Jesus from the dead dwell in you, he that raised up Christ from the dead shall also quicken your mortal bodies by his Spirit that dwelleth in you"*. The quickening brings acceleration. It stimulates and revives.

Ephesians 2:1-6 declares, *"[1]And you hath he quickened, who were dead in trespasses and sins; [2]Wherein in time past ye walked according to the course of this world, according to the prince of the power of the air, the spirit that now worketh in the children of disobedience: [3]Among whom also we all had our conversation in times past in the lusts of our flesh, fulfilling the desires of the flesh and of the mind; and were by nature the children of wrath, even as others. [4]But God, who is rich in mercy, for his great love wherewith he loved us, [5]Even when we were dead in sins, hath quickened us together with Christ, (by grace ye are saved;) [6]And hath raised us up together, and made us sit together in heavenly places in Christ Jesus"*.

We are alive in Christ Jesus because we have been spiritually revived. Glory to God! At the appearance of Jesus Christ (at the last trumpet), we will experience a resurrection. The Bible says, *"[16]For* *the Lord himself shall descend from heaven with a shout, with the voice of the archangel, and with the trump of God: and the dead in Christ shall rise first: ^{17}Then we which are alive and remain shall be caught up together with them in the clouds, to meet the Lord in the air: and so shall we ever be with the Lord"* (1 Thessalonians 4:16-17 KJV).

In conclusion, the concept of a reboot (in our lives and in our faith) offers a profound opportunity for renewal and transformation. Just as a computer requires a restart to clear out errors and improve performance, we, too, must periodically reset to realign ourselves with our divine purpose and potential. Embracing this reboot means allowing God to renew, restore, rename, and revive us. We must shed old ways of thinking and functioning to make way for a fresh start.

As we journey through life, we will inevitably face challenges and moments of stagnation. However, by turning to the Lord, we can find the strength and guidance to navigate these difficulties. In Hosea 6:1-2 (NIV), God reminds us of His promise to heal and restore us. The scripture says, *"^{1}Come, let us return to the Lord. He has torn us to pieces but he will heal us; he has injured us but he will bind up our wounds. ^{2}After two days he will revive us; on the third day he will restore us, that we may live in his presence"*. Let us wholeheartedly embrace this divine reboot. We must allow it to transform our lives and lead us into a future filled with hope and purpose.

Inspirational Downloads

Embracing Change

-7-
FROM STRUGGLE TO STRENGTH
(Navigating the Path of Recovery)

> **Recovery**
>
> Imagine sitting down at your computer, ready to finish an important presentation. You attempt to open the file, but it's not there. Panic starts to rise as you frantically search through every folder and drive. With growing horror, you realize that all your important documents and cherished photos are gone. The thought of losing irreplaceable memories and crucial files feels overwhelming. Desperate, you grab your phone and call a technical specialist (your voice trembling as you explain the situation). With your heart pounding, you rush to the shop, clinging to a fragile thread of hope.
>
> The technician connects your laptop to his recovery tools, and you anxiously watch every move. After what feels like an eternity, the technician smiles and says, "I found them". Relief floods through you as you cry tears of gratitude. Your precious memories and critical files are safe. There is great joy when we recover what was lost. Just as we feel immense relief when we regain something precious, God wants to recover and restore the lost pieces of our lives. He wants to bring us back to wholeness and joy.

In cybersecurity, the concept of "recovery" mirrors an ancient tale of resilience. In this chapter, we will embark on a fascinating exploration of recovery in the digital realm. We will also draw parallels from an extraordinary journey of "recovering all" from loss and adversity.

As cyber threats continue to escalate in complexity and frequency, organizations and individuals are faced with the inevitable challenge of securing their digital assets. Therefore, recovering from cyber-attacks and data breaches requires a strategic approach that goes beyond mere prevention. It must encompass a robust response and restoration strategies. In the face of adversity and challenges, the journey of recovery unveils a profound opportunity for growth and restoration. In this chapter, we will explore a powerful tale of resilience and redemption that echoes through the ages. It is the biblical account of David. When he faced a devastating loss and was told to "recover all", he embarked on a transformative quest that required urgency. It also illustrated the potential (within us all) to reclaim what was lost and emerge stronger than ever before. Let us draw inspiration from David's unwavering determination and faith. We will also discover how this ancient story continues to hold timeless relevance in navigating the depths of our own challenges and triumphs.

Recovery is a return to one's normal state of health, mind, and strength. It is also the action or process of regaining possession or control of something stolen or lost. In Cyber Security, the main focus is on information and how to secure it. If the information is lost or stolen, how can it be recovered? It is pertinent that a recovery plan (or backup plan) is in place. It should include cyber recovery and disaster recovery strategies. Data backups should be secured onsite and offsite. A recovery plan should involve activities (after an incident or event) which restore essential services and operations. Recovery can range from short term to long term where capabilities are fully restored.

Considering the risks involved in both cyber security and in our lives, everyone should have a recovery plan in place if something happens. A biblical perspective on this can be found in 1 Samuel 30:1-20 (KJV). Let's look at it together.

Ask yourself this question: *What would I do after an attack has happened and I discovered that what is most important to me was taken?* David and his men had to face this tragic situation. The Bible says that they wept loud until they had no strength left to weep. If they stayed there and continued weeping over what transpired, nothing would have changed the outcome of the situation. Remember: It is ok to take a moment and weep over your loss. However, there must come a point where you wipe those tears away and do something about your situation.

When David and his men suffered this difficult attack from their enemies (the Amalekites), they lost their wives, children, and valuable property. Nonetheless, they had to get up and do something about it. Recovery does not just happen. It requires planning and work for recovery to take place.

David was not just faced with the issue of his personal loss. He was also faced with distress from the men, who were planning to stone him because of their loss. What an adverse condition David was in! What could be done to get out of such a situation? Have you ever been faced with the loss of your spouse, children, or your support system? What would you have done if you were faced with the same situation David was in?

Despite the loss, distress, and bitter resentment towards David, the Bible says that he encouraged himself in the LORD (1 Samuel 30:6-8 KJV). This is the first key to your recovery. The source of your encouragement comes from the LORD. You must find inspiration in adverse conditions. What better way is there than by turning to

God? Encouragement enables you to regain focus. You need to focus on knowing what to do next. You have experienced loss, and you need an action plan.

Obtaining an action plan (with clear and concise instructions) is the second key to your recovery. In 1 Samuel 30:8, you discover that David enquired of the LORD saying, *"Shall I pursue after this troop? Will I overtake them?"* This was a great inquiry that David presented to God. When you are uncertain of what should be done during an attack and you experience loss, you must have a strategic plan for recovery. The plan must provide a means of getting back what was stolen and fully regaining what you had before the attack. I submit to you this thought, never leave God out of the plan. God, who knows the end from the beginning, can give insight to what you do not know and what you need to expect during the recovery process.

David asked God what he should do next. As a result, the LORD answered him saying, *"Pursue, for you shall surely overtake them and without fail recover all."* (1 Samuel 30:8 NKJV). This is a shouting moment right here. Glory to God! To pursue is to engage in a course of action. It is time to put the plan into full effect. It is time to do it! This is the third key to your recovery plan.

Work that strategy out in compliance with what it requires. Follow through on what the plan calls for. When David and his men pursued, it was in obedience to what God's strategy was, in order for them to see full recovery. As you continue reading through 1 Samuel 30, you will discover David's pursuit for recovery. However, I would like to highlight these two passages of scripture that provide what happened according to what the LORD told David.

In 1 Samuel 30:18-19 (KJV), the Bible says, *"ⁱ⁸ And David <u>recovered all</u> that the Amalekites had carried away: and David rescued his two wives. ¹⁹ And there was nothing lacking to them, neither small nor great, neither sons nor daughters, neither spoil, nor any thing that they had taken to them: <u>David recovered all</u>"*. A full recovery took place for David and his men by following God's recovery plan.

David chose to stay focused on the real issue and not be distracted. Therefore, he encouraged himself in the LORD. Afterwards, he inquired of the LORD to receive a plan of action from God. Finally, he obeyed the divine strategy. David received instructions, pursued his enemies, and recovered everything they stole.

If you are believing God for recovery today, remember David's model and allow the LORD to help you recover all. If you follow God's plan you shall see a full recovery! You shall see recovery in your health, marriage, finances, business, mental capacity, career, family relationships, and in your ministry in the name of Jesus Christ. Amen!

Inspirational Downloads

-8-
DIVINE FIREWALL
(Safeguarded by God)

> **Firewall**
>
> Imagine it's crunch time. You're working on an important project, deep into your flow, when suddenly, your screen flickers. Pop-ups start appearing out of nowhere, and files begin disappearing before your eyes. You try to close the windows, but it's too late—your system has been compromised. The realization hits you like a wave: your data, your work, everything is vulnerable. In a state of panic, you reach out to cybersecurity experts. As they examine your system, they explain how easily this could have been avoided. "You didn't have a firewall," they say. Without it, your computer was like an open door, inviting threats in. This experience teaches you the hard way that a firewall isn't just an option - it's essential. It's the barrier between your valuable data and the dangers lurking online, a crucial line of defense in a world where threats are constant and relentless. Just as a firewall protects your computer from harmful intrusions, God wants to be the firewall for our lives, shielding us from dangers we often don't even see coming.

In today's interconnected digital age, the concept of a firewall holds profound relevance to safeguarding against security threats. It is similar to the protective measures of the whole armor of God described in Ephesians 6:11-18. Zechariah 2:5 (NIV) states, *"And I myself will be a wall of fire around it, declares the Lord, and I will be its glory within"*. God promised to be a firewall protecting Jerusalem, a city without physical walls due to its vast population and livestock. A firewall acts as a barrier to unauthorized access in the realm of information security. In the same manner, the whole armor of God serves as a spiritual defense and safeguard (for the Believer) against the schemes of the devil, principalities, powers, rulers of darkness, and spiritual wickedness in high places.

The helmet of salvation protects the mind; the breastplate of righteousness guards the heart; and the shield of faith deflects spiritual attacks, much like how a firewall blocks malicious data packets. Understanding these parallels underscores the importance of standing strong with God's armor and protection. It is His armor and protection that ensure our spiritual and digital environments remain secure and fortified against unseen adversaries, who seek unauthorized access to us.

Before we dive into the spiritual significance of divine protection, let us look at what a firewall is and how it protects networks and data. A firewall is a critical component in the realm of information security. It acts as a barrier between a trusted internal network and an untrusted external network (e.g. the internet). It is designed to monitor and control incoming and outgoing network traffic based on predetermined security rules. Firewalls can be implemented as hardware, software, or a combination of both. By analyzing data packets and determining whether they should be allowed or blocked, firewalls help prevent unauthorized access, cyberattacks, and other security threats.

Firewalls serve as the first line of defense in protecting computer systems and networks from malicious activities. They achieve this by enforcing strict access controls, which allow legitimate traffic, while blocking potentially harmful data. For example, a firewall can be set to block access to certain websites, restrict the use of specific applications, or deny connections from suspicious IP addresses. This selective filtering ensures that only safe and authorized communications are permitted. In return, the risk of data breaches and other cyber threats is significantly reduced.

One of the primary functions of a firewall is to prevent unauthorized access to private networks. By examining the source and destination of each data packet, firewalls can identify and block unauthorized attempts to gain entry. This is particularly important for protecting sensitive information, such as personal data, financial records, and intellectual property. Additionally, firewalls can be configured to provide logging and alerting capabilities. This would allow network administrators to monitor and respond to potential security incidents in real-time.

In addition to preventing unauthorized access to private networks, firewalls play another crucial role. They prevent the spread of malware and other malicious software. This task is accomplished by inspecting incoming traffic for known threats and suspicious patterns. During this time, firewalls can detect and block malware before it infiltrates the network. This proactive approach helps maintain the integrity and availability of network resources. It also ensures that systems remain operational and secure. Overall, firewalls are essential to protecting digital assets and maintaining a secure and trustworthy, computing environment.

Similar to the protection and capabilities provided by firewalls, the whole armor of God serves as the ongoing protection for Believers. God's protection empowers us to stand strong in the earth. Keep in mind that to stand is not to merely be in an upright position with all weight on your feet. Instead, it embodies a deeper concept of maintaining steadfastness and endurance in the face of challenges. In other words, we must occupy a position that is unyielding. Our position must also be rooted in truth, righteousness, peace, and prayer. These safeguards serve as pillars of reinforcement, which ensure that we do not falter or fail.

The SAFEGUARD of TRUTH

Let us explore the first safeguard: TRUTH. Ephesians 6:14 (KJV) instructs us to, *"Stand therefore, having your loins girt about with truth..."*. Biblically speaking, the concept of girding one's loins with truth involves being securely fastened or bound by the truth of God's Word.

John 17:17 emphasizes to, *"Sanctify them through thy truth: thy word is truth"*. The Word of God is not just essential for our spiritual sustenance. It is also transformative and purifying when internalized. Just as food nourishes the body, the Word nourishes the spirit and cleanses the heart. This process purges us of hidden sins and prepares us for greater fruitfulness in our walk with Christ. Psalm 119:11 reinforces this truth. Let us read it together: *"Thy word have I hid in mine heart, that I might not sin against thee"*. Embracing and obeying God's Word not only aligns us with His will. It also shields us from sin and empowers us to stand firm in our faith journey. This journey is rooted in our belief that Jesus Christ is our Lord and Savior.

Everyone, who believes in Jesus Christ and loves Him by receiving His Word, will experience His life. John 7:38 (KJV) declares, *"He that believeth on me, as the scripture hath said, out of his belly shall flow rivers of living water."* Life-giving streams flow from our lives because of His Word. As we embrace the Word of the LORD, a transformation takes place within us. Our will aligns with God's will. Our minds are renewed, and we are revived by His Truth. The Word of God enlightens, sanctifies, and strengthens us, as it is the very essence of Truth. This Truth revitalizes us and equips us to fight, finish the race, take flight and keep the faith.

TRUTH Revives Us for the FIGHT
Ephesians 6:14-18 describes the armor of God (our defense) against the schemes, strategies, and deceits of the devil. The scripture reveals that God's primary defense against lies and deception is the TRUTH. Each part of the armor serves a vital purpose in the believer's life. It also ensures that our engagement in spiritual warfare is approached with divine strategies. Therefore, we must be suited up, *"having our loins girt about with truth"*, in order to stand firm against the enemy's cunning tactics.

The devil is so deceptive that he can manipulate people into believing lies, as though the lies were the actual truth. The devil is slick and crafty, which makes it imperative for us to be prepared for battle. By wearing our belt of TRUTH, we are equipped to combat every false belief and manipulation.
Truth is the strength of our loins. It holds us together during adversity. Just as soldiers gird their loins with a belt to secure their tunics for battle, the truth holds us together and guarantees we do not lose heart. Daily, we must resist the devil and know that the Word of God will revive us for the fight. As good soldiers of Jesus Christ, we must adhere to 1 Peter 5:8 (KJV), which says, *"Be sober, be vigilant; because your adversary the devil, as a roaring lion, walketh about, seeking whom he may devour"*.

TRUTH Revives Us for the FINISH
We are not only engaged in a fight, but we are also in a race. In order to complete this race, we need the endurance and strength that God provides through His word. He sends His word to revive and strengthen us so that we can finish strong. 1 Corinthians 9:24

(NIV) urges us to *"run in such a way as to get the prize"*. However, we will never finish the race and win the prize if we are weighed down by excess baggage. A runner wearies quickly if burdened by heavy clothing or shoes. Because *"The King's business requires haste"* (1 Samuel 21:8 KJV), we must lighten our load to swiftly do what the LORD requires.

In Galatians 5:7 (NKJV), the Apostle Paul wrote, *"You ran well. Who hindered you from obeying the truth?"* When our pace slows down, we must recognize WHO is hindering us. Often, it is external influences that weigh us down. It is at this time that we need to identify and shed the baggage that comes from these ties.

Let us reflect on the concept of a firewall, which blocks external sources from gaining unauthorized access to internal information. Once we have identified the external influences that affect us internally, we must then set boundaries and rules to blockade our hearts and our thoughts. The Apostle Paul declared that he had not run in vain but finished the course. By blocking our hindrances, releasing them, and focusing on the race set before us, we, too, will run well and finish the course.

TRUTH Revives Us for FLIGHT
Isaiah 40:31 (KJV) states, *"But they that wait upon the Lord shall renew their strength; they shall mount up with wings as eagles; they shall run, and not be weary; and they shall walk, and not faint"*. This renewal is essential for our spiritual journey. It will give us the stamina to press on.

The eagle sheds its old feathers and receives a new covering, which symbolizes renewed life. Similarly, when we wait on God, we are renewed. As believers, living a spiritual life is like taking flight. However, many of us are afraid of heights, too earthbound, and lost in the fog of doubt and despair. We are submerged in limiting concepts and ideologies. Brothers and Sisters, it is time for these limitations to be broken so that we can fly high. God's TRUTH is the fuel that we need. His word will revive, sustain, and empower us for the journey. It will also enable us to soar above difficulties, disappointments, trials, and failures, similar to how an eagle soars high in the sky. When we rise on the wings of TRUTH, we gain a clearer perspective and an advantageous position.

TRUTH Revives Our FAITH

Romans 10:17 (NKJV) affirms, *"So then faith comes by hearing, and hearing by the word of God"*. The Word of God breathes life into our faith and fortifies us to stand firm in our beliefs. Each encounter with God's word should ignite and replenish our faith. We must be open vessels that are ready to receive God's word with joy and allow it to fill us with faith. Psalm 119:28 (NKJV) declares, *"...Strengthen me according to Your word"*. This scripture emphasizes how the Word of the Lord encourages and empowers us to be strong.

Daily, we require spiritual nourishment and divine revelation to continually build and sustain our faith. It is imperative to seek God's presence earnestly. Today, more than ever, God's people need to be fortified on the foundation of their most holy faith. Jude 1:20-21 (NKJV) exhorts, *"But you, beloved, building yourselves up on your most holy faith, praying in the Holy Spirit, keep yourselves in the love of God, looking for the mercy of our Lord Jesus Christ unto eternal life"*.

Faith is crucial for us as believers. Hebrews 11:6 emphasizes, "But without faith it is impossible to please Him, for he who comes to

God must believe that He is, and that He is a rewarder of those who diligently seek Him." Additionally, 1 John 5:4 declares, "For whatever is born of God overcomes the world. And this is the victory that has overcome the world, even our faith." Faith not only pleases God but also ensures victory in our lives. May our faith be revived, empowering us to triumph in every aspect of our journey.

The SAFEGUARD of RIGHTEOUSNESS

We have explored the first safeguard: TRUTH. Our divine protection and strength come from truth, which shields us against the lies and deceptions of the enemy. The second critical safeguard for standing strong is: RIGHTEOUSNESS. Ephesians 6:14 (NKJV) continues to instruct us in *"...having put on the breastplate of righteousness"*.

The breastplate of righteousness is profoundly significant. Just as a physical breastplate protects the heart from harm in battle, spiritual righteousness safeguards our hearts from sin and deception. It is a standard every believer must uphold, as it preserves the core of our thoughts and emotions. In ancient times, the High Priest's breastplate bore the names of the twelve tribes of Israel, symbolizing their closeness to God's heart. Today, it signifies God's desire to govern our lives from within our hearts. Psalm 37:23 (NKJV) assures us, *"The steps of a good man are ordered by the Lord: and he delights in his way"*.

In addition, the breastplate contained the names of Abraham, Isaac, and Jacob. It also contained the words *Shivtei Yeshurun* (tribes of Jeshurun) and featured all twenty-two letters of the Hebrew alphabet. A piece of parchment (known as the Urim Vetumim) was inserted within the folds of the breastplate and it contained the Tetragrammaton, God's sacred name. As Abraham,

Isaac, and Jacob (patriarchs of the faith) were significant to the tribes of Israel, Jesus Christ is also significant to us as Believers. As a result, we have assurance and security in the name of Jesus Christ. His name is near and dear to our hearts. It is precious and we remain steadfast in our faith in Him.

Furthermore, the breastplate of the High Priest (adorned with twelve precious stones and divine inscriptions) served as a guiding garment. It wasn't merely symbolic, but it was practical in connecting God intimately with His people. Similarly, our righteousness connects us to God's divine guidance and protection. As we walk in righteousness, guided by the Holy Ghost, we renounce deceit and uphold the truth of God's Word. This ministry of the Spirit leads us into all truth, empowering us to preach the gospel boldly and bring light to those lost in darkness (2 Corinthians 4:1-10). Through this divine connection, the light of the glorious gospel of Christ shines in our hearts and transforms us into His image by the Spirit of the Lord. This transformation influences our morality, mindset, and disposition. It also aligns us more closely with God's will and purpose.

The SAFEGUARD of PEACE

We have explored two safeguards thus far: TRUTH and RIGHTEOUSNESS. The third safeguard we will discover is: PEACE. Ephesians 6:15 (NIV) says, *"And with your feet fitted with the readiness that comes from the gospel of peace."* In a world full of uncertainties and challenges (i.e. health problems, financial strain, mental distress, loss of loved ones, and attacks from adversarial forces), we need the peace of God more than ever. Romans 15:13 tells us that peace comes from the Holy Ghost. Righteousness, peace, and joy are in the Holy Ghost, which embodies the experience and operation of the Kingdom of God.

This divine peace is a shield against the trials we face. It provides a sanctuary amid life's storms.

Mark 4:35-41 tells us that when a storm threatened the disciples' ship, Jesus (asleep amidst the chaos) was awakened by His frightened disciples. He rebuked the storm by saying, *Peace, be still*. This demonstrated Jesus' authority as the Prince of Peace. It also teaches us that peace can be declared to calm any storm in our lives. As Believers, we must speak by the Spirit and declare peace to settle the atmosphere around us, no matter the challenges we face. The authority of Jesus to command peace in the storm demonstrates that through faith and the Holy Ghost, we, too, can find calm in chaos.

In English, peace is defined as the absence of civil disturbance or hostilities and a state free from internal and external strife (Baker's Evangelical Dictionary of Theology, 1996). However, in Hebrew, peace "Shalom" is defined as harmony, wholeness, right relationships, success, prosperity, and victory over opponents. When we use "Shalom" as a blessing, we invoke God's perfect peace. This covenant of peace is sealed by God's presence and assured to us by the Holy Ghost. It is this profound, holistic peace that sustains us, as it encompasses all aspects of our well-being (physical, emotional, and spiritual).

Isaiah 26:3 (KJV) declares, *"Thou wilt keep him in perfect peace, whose mind is stayed on thee: because he trusteth in thee."* Our minds must align with God and trust Him for true peace. The THINKER must align his or her thoughts on God and TRUST Him. This alignment with God's will and trust in His power are essential to experiencing His perfect peace. The peace of God is not just a momentary calm. It is a deep, enduring state of well-being that surpasses human understanding. It guards our hearts and minds, and it enables us to remain steadfast in faith despite life's uncertainties.

Jesus promised His peace in John 14:26-27 (KJV), which states, *"²⁶But the Comforter, which is the Holy Ghost, whom the Father will send in my name, he shall teach you all things, and bring all things to your remembrance, whatsoever I have said unto you. ²⁷Peace I leave with you, my peace I give unto you: not as the world giveth, give I unto you. Let not your heart be troubled, neither let it be afraid."* Through His sacrifice, Jesus established peace between God and man, breaking down barriers and reconciling us to God. This peace, brought to us by the Holy Ghost, guards our hearts and minds. It also allows us to live confidently, despite external trials. This divine peace is distinct from the peace the world offers. It is a peace that transcends circumstances and fills our hearts with unwavering assurance and joy.

Ephesians 2:14, 16 explains how Jesus established peace through His sacrifice. The scripture states, *"¹⁴For he is our peace, who hath made both one, and hath broken down the middle wall of partition between us; ¹⁶And that he might reconcile both unto God in one body by the cross, having slain the enmity thereby."* This reconciliation through Christ ensures that we are no longer strangers but fellow citizens with the saints and members of God's household.

God's peace is our inheritance as believers. It is a constant presence that empowers us to face life's challenges with courage and hope. As stated in Philippians 4:7 (NIV), *"And the peace of God, which transcends all understanding, will guard your hearts and your minds in Christ Jesus"*. God's peace will also help us to think on things that are true, noble, right, pure, lovely, and admirable (Philippians 4:8 NIV). This divine peace anchors us and enables us to navigate life's storms with a steadfast heart that is rooted in the love and power of Christ.

The SAFEGUARD of PRAYER

As we embrace the divine peace that guards our hearts and minds, we turn our attention to the fourth and final crucial safeguard in our spiritual journey: The power of PRAYER. Ephesians 6:18 (KJV) mentions *"Praying always with all prayer and supplication in the Spirit, and watching thereunto with all perseverance and supplication for all saints;"*

The power of prayer ties in with the firewall of a network in several meaningful ways. Just as a firewall serves as a barrier to protect a network from unauthorized access and potential threats, prayer acts as a spiritual defense. This defense safeguards believers from negative influences and attacks. See the following parallels:

1. **Protection and Security**: A firewall monitors incoming and outgoing network traffic (based on pre-determined security rules) and ensures that harmful data does not penetrate the network. Similarly, prayer acts as a spiritual shield and invokes divine protection against spiritual dangers and harmful influences. Ephesians 6:18 (NIV) says, *"And pray in the Spirit on all occasions with all kinds of prayers and requests. With this in mind, be alert and always keep on praying for all the Lord's people."* The Lord will rescue us from every evil attack and bring us safely to His heavenly kingdom.

2. **Constant Vigilance**: Firewalls protect the network at all times. When we practice prayer consistently, it makes us spiritually vigilant and connected to God. He, in turn, offers us constant protection and guidance. 1 Peter 4:7 encourages us by saying, *"The end of all things is near. Therefore, be alert and of sober mind so that you may pray"*.

3. **Filtering and Control**: Firewalls filter out unwanted traffic, allowing only safe and authorized data to pass through. Prayer helps believers discern and filter out negative thoughts, temptations, and influences so that our lives may align with God's will. Philippians 4:6-7 (NIV) says, *"⁶Do not be anxious about anything, but in every situation, by prayer and petition, with thanksgiving, present your requests to God. ⁷And the peace of God, which transcends all understanding, will guard your hearts and your minds in Christ Jesus."*

4. **Establishing Boundaries**: Firewalls establish clear boundaries between the internal network and external threats. Prayer helps believers establish spiritual boundaries, which reinforces their faith and moral values against external pressures and temptations. When we pray, it must be done in faith. We must believe that whatever we have asked according to God's will, it is ours. In Mark 11:24-25 (NIV), Jesus showed that prayer is done through faith and forgiveness.

5. **Updates and Adaptation**: Firewalls require regular updates, in order to adapt to new threats. Similarly, regular, earnest prayer helps believers stay spiritually updated and adaptable. This, in turn, enables believers to seek wisdom

and strength to face new challenges and changes in their lives. Romans 12:12 (NIV) encourages us to "Be joyful in hope, patient in affliction, faithful in prayer". This practice emphasizes the ongoing need for spiritual renewal and resilience.

By now, we can see that prayer (much like a firewall) is a crucial safeguard. It protects, guides, and empowers believers. It also reinforces their spiritual resilience and connection to God.

In conclusion, the four safeguards (TRUTH, RIGHTEOUSNESS, PEACE, and PRAYER) serve as essential elements of the armor of God. Each one plays a vital role in our spiritual defense. TRUTH (our foundation) protects us against the lies and deception of the enemy. It also enables us to stand firm. RIGHTEOUSNESS (symbolized by the breastplate) guards our hearts against sin and moral compromise. It also ensures that we walk in alignment with God's will. PEACE (brought by the Holy Ghost) calms the storms of life and provides a sense of stability and assurance, even amidst chaos. Finally, PRAYER (our direct line of communication with God) empowers us to seek His guidance and strength. It also fortifies our resolve and keeps us spiritually alert.

Firewalls filter out harmful traffic to protect a network from external threats. Similarly, the armor of God shields us from the spiritual attacks and challenges we face in our Faith journey. Each piece of this divine armor is designed to: (1) protect specific aspects of our spiritual well-being and (2) ensure that we remain resilient and steadfast.

Truth acts as a filter against deception; Righteousness acts as a barrier against moral corruption; Peace acts as a buffer against turmoil; and Prayer acts as a constant connection to divine support. Together, these safeguards create a comprehensive defense system that empowers us to stand strong and victorious. As a result, we are fortified by the power of the Holy Ghost and ready to face any challenge that comes our way.

My prayer for us today is that God will protect us from all emerging threats. I pray that we will all have secure relationships with Him and others. May we experience the necessary upgrades that will allow us to stand resilient for our future. It is our time to be renewed and refreshed as we stand strong, having our hearts fortified against all evil. Let us remember that we are secured in Jesus Christ, who is the author and the finisher of our faith. May God bless us, preserve us and cause us to experience the transforming power of His love!

Inspirational Downloads

www.ingramcontent.com/pod-product-compliance
Lightning Source LLC
Chambersburg PA
CBHW070532100426
42743CB00010B/2049